The Gardener's Chairside Reader

Bud Brinkley

The Gardener's Chairside Reader

Flying V Publishing
www.budbrinkley.com

The Gardener's Chairside Reader Copyright © 2020 Flying V Publishing

All rights reserved. No part of this book may be reproduced in any form or by any electronic or mechanical means, including information storage and retrieval systems, without written permission from the author, except in the case of a reviewer, who may quote brief passages embodied in critical articles or in a review.

Published by Flying V Publishing
Texas USA

ISBN: 978-0-578-75741-4

A great companion to have while sitting next to a cozy fire and dreaming of gardening! *The Gardener's Chairside Reader* is a collection of short stories, tips, gardening wisdom, and advice. Bud has years of gardening experience and his sometimes humorous views of country life make for entertaining as well as informative reading. Trying out the latest gardening ideas and techniques while stubbornly holding on to the traditions of the past has created some interesting observations that will keep you smiling and saying, "Maybe I should try that too!" Bud's childhood memories and stories are sprinkled throughout, along with a few spine-chilling descriptions of days gone wrong. Throw in some useful tips and advice and you have a book to settle in with while drinking a cup of coffee each day!

From the Readers...

" A funny and revealing look at real country life"
R. Taylor

"A daily dose of garden help"
Lydia Jameson

"It's like sitting on the front porch talking to a farmer each day!"
Cathy Broussard

Dedication

To Laura Beth for her unending encouragement.

Acknowledgments

My family, garden, and dogs have suffered through my bouts of daily temper tantrums, sulkiness, and moments of complete indifference, confusion, hair loss, sleepless nights, and lack of personal hygiene, so that this book could be completed.
You have my sincere thanks and gratitude.

Also by Flying V Publishing

Building a Wooden Jon Boat
Tips, Techniques, and Ideas

Coming Soon!
The Tightwad Gardener

Table of Contents

Introduction   XVII

My Garden Shed   19

Charlotte's Web   23

A Super-Duper Garden Seeder   27

My Failure With Drip Irrigation   31

Morning Coffee and Blessings   35

A Twisted Plot   39

Tomato Temptations   43

Gardening From the Comfort Of Your Own Bed! Click Now!   47

Tractor Commercials   51

Water Hose Kinks and Thrills   55

Till vs. No-Till vs. My Aching Back   59

Cover Crops   63

Iron Relics From the Past   67

The Feedstore Red Light District   71

Contents

Where Did All Of the Jars Go?   75

Hoe-Down!   79

Crooked Rows, Moses, and Pearl Beer   81

The Wormville Motel   85

Life in Three Days   89

Green Acres   93

A Day In the Sunshine   97

It's A Tilthy Business   99

Wheel Hoe Wisdom   103

Hay Balers and Lands' End   105

The Brinkley Shell-O-Matic   109

Lunar Post Holes   113

Christmas Porkchop   115

Farm Gate Envy   117

Salt Of the Earth   121

A Quiet Morning With Big Bertha   125

The Florida Weave   129

Contents

Three Dot Snuff, Cigars, and Old Dogs   131

Suspended Animation   135

A Honeybee's Quest   139

Grace   143

Introduction

One afternoon, while sitting in my garden literally watching the paint dry on my new garden shed, the idea to write this book came into being. Having written a few previous books on varied topics, paired with my keen interest and considerable experience with gardening, the idea naturally progressed. While discussing this proposal with my sweet wife over a glass of iced tea (it was a very hot day), she suggested that I also include some of the unique experiences I'd encountered while growing up in the country. You see, I was raised on a farm beside a dirt road near a small town in Texas. Most of my earliest memories stem from this wonderful place. I was blessed with long summers playing in the fields and pine forests of the area. Watching my folks deal with the quirks and curiosities of country life forever influenced me. Those days have long since passed, yet I still find myself drawn to the same old ways and traditions of the past.

My intent with this book is to give my readers a few minutes of reflection, a bit of useful information, and perhaps some entertainment each day. Hopefully, you will take some time to sit down and steal away a few quiet moments with this book and will enjoy what you read. I have throughly enjoyed writing this book and I appreciate your interest and participation. ...Besides, since my retirement, watching paint dry and drinking iced tea... what else do I have to do?

Bud

My Garden Shed

Have you ever had an original idea that grew so large that you laugh at yourself for even trying it in the first place? Here is a case in point; I have had several gardens in several different places on my ranch at various times throughout the years and this year I decided to move it yet again in order to find the perfect spot for my new spring garden. I picked a nice plot near what used to be an old homestead on our property and commenced to breaking ground, completely unaware that my plans were about to spiral out of control.

Since this particular area did not have a water supply nearby, I knew I had to install a waterline from the well at my house down to this new garden area. It was about a 350ft run and manually digging a ditch that length was out of the question, so I used a tractor and a middle buster plow to dig the trench. Just the digging part with the tractor took several days. I drove to town to get loads of PVC pipe for the waterline, glancing woefully each time as I passed by my old previous garden area which already had a perfectly usable water supply. I should have just kept my garden there, but no, once again I am determined to build the Taj Mahal of gardens. Nevertheless, I soldiered on in the scalding sun,

Bud Brinkley

digging ditches, gluing pipe sections together, and setting up a new water spigot on a post next to my garden. After connecting the pipe and then covering up all 350ft. of ditch with soil, I now finally had water in my new garden at last! I had yet to plant a single seed mind you, as I was distracted for a week messing around with this task.

No sooner had I planted some new seeds and begin the actual process of gardening, another hair-brained idea popped into my head. It was hot as you know what and the old Pecan tree next to the garden wasn't leafed out yet so it wasn't giving me much shade. Also, I had no comfortable place to sit down and rest. I looked out across the pasture to an old shed in the distance that I had built years ago. In a heat-induced moment of quick decision, I decided I needed that shed to be here at my new garden site. This old shed was really just a rusty tin roof supported on 4 poles with no walls or any other structure. No matter, I eagerly devised a plan to cut the poles with a chainsaw and lower the shed to the ground using a tractor and then move the whole contraption to it's new location.

I enlisted the help of my father-in-law and somehow we managed to get the old shed moved and mounted up high on 4 new poles cemented into the ground. I had it located in a suitable spot near the big pecan tree right next to my new garden plot. I was congratulating myself on completing the garden area finally and getting back to work tending to the young plants that were by now sprouting. But wait, suddenly I had yet another hair-brained idea. I could build out the shed with some walls and a floor and maybe even an old farm sink to make a super-duper garden shed to process and wash vegetables! Again, I set out for several trips to town to purchase big loads of lumber and each trip, I was passing by my old garden spot wondering why would I make so much extra work for myself. To add a third hair-brained idea on top of the last two, I decided to build what had by now evolved into a full sized tiny-cabin in the style often seen in the Texas Hill Country. I always admired the designs of the early settler cabins built by the first German immigrants in Texas. Their cabins usually had steep roofs and large windows with board and batten siding. I decided to build my cabin the

The Gardener's Chairside Reader

same way. Now here I was, fully involved into a major building project including architectural research, design and landscaping.

My poor garden was still growing but it was tough to keep building each day alone while keeping a newly planted garden on track also. Eventually, I completed the tiny cabin and I even built a nice rustic front porch on it so I could sit and rest in the shade and look out onto my beautiful garden. Finally, I managed to get back to the original task of actually growing vegetables. There is just one nagging little problem that is bugging me though. My garden cabin turned out much nicer than I had originally anticipated, but it's a bit too small. I am already devising plans of expanding it to include a full size kitchen, wood heater, and a nice sleeping area, and…

It's pretty easy to get sidetracked when you just need some shade and a place to sit and rest …

Bud Brinkley

The Gardener's Chairside Reader

Charlotte's Web

I use a Kawasaki Mule 4-wheeler to get around on my ranch. I am riding in it sometimes all day on busy days, and most days I use it to haul around cargo where it's needed. This Mule and I have a long history together and we are both about worn out but that is a subject for another story. This story is about a freeloading passenger.

I noticed one morning as I was climbing in the seat that a beautiful and delicate spider web had been built just ahead of the steering wheel. It was a dewy springtime Texas morning and the spider web was twinkling with drops of moisture in the sunlight. The web was about 6 inches in span and culminated into a neatly woven tunnel that led into a hole in the dashboard that enclosed the steering wheel shaft. I did not destroy the web as frankly I enjoyed it's careful design and construction and anyway, I was in a hurry to get to work in the garden.

Historically, I have a solidified hate relationship with 3 creatures in this world - Snakes, Spiders, and Scorpions. They all receive a swift death in my presence, but hey that's just me I suppose. I was bitten once by a Brown Recluse spider when I was a young man and I came very

Bud Brinkley

close to dying from that bite. I still vividly remember how sick I was for many weeks before slowly recovering. Thus, I have always squashed any spiders I see in the house. One thing I have noticed about myself as I grow into old age is that I have become more of an old softy these days and sometimes think twice about killing a creature, especially if it's outside in it's home environment and not really hurting anything.

Anyway, back to the subject of the spider web on the Mule. About lunch time I was getting ready to ride back to the house and I walked up and saw the spider that had so carefully built it's new web crawling around on my dashboard. It was fairly large but it didn't look particularly deadly to me. When it saw me, it immediately tucked back into it's tunnel behind the dashboard. As I was rumbling along back to the house I could just see it peeking out at me riding along with me. That evening, sure enough I saw the spider again as we went for a ride to a far section of the ranch after supper. I sat there smoking my evening cigar and watched the curious spider slowly crawl out and get a good look at me. At that point I started talking to her. Yes, I decided it was a she and I named her Charlotte. (Get it?) I said, "As long as you don't mess with me, I won't mess with you." I also told her that she was in for a bumpy ride sometimes and to hang on good and don't go around jumping out and scaring the hell out of me and causing me to surely wreck the Mule.

Our mutual agreement seemed to work well and for the next month or so I became somewhat attached to Charlotte the spider. Each morning I would notice that she had repaired any damage to her web. I could tell that she had been out all night weaving and repairing and catching bugs and such. She would ride with me each morning bumping along our old road as I took the trash out or went down to the mailbox. In fact Charlotte and I traveled quite a few miles together around the ranch in perfect harmony. It was about as close as a man could get to a spider I suppose and I like to think that we enjoyed each other's company.

The Gardener's Chairside Reader

One morning I noticed her web wasn't repaired as neatly as usual. I looked and saw her tucked away in the tunnel so I wasn't too concerned at first. Then each day her web became a little more tattered and I could tell that she wasn't catching very many bugs anymore. I even went so far as to catch a small grasshopper myself and put it into her web one evening at dark. I told you I was becoming an old softy…The next day the grasshopper was untouched and I knew what it meant. I could not find Charlotte in the dashboard and I was fairly certain she was finished with her short life. Still I left the old tattered web in place behind the steering wheel just in case she returned home. It's been about two weeks now and I guess I'll go outside and clean off the old web and wipe down the dashboard because my little friend has gone. We had some great times doing our daily chores together. R.I.P. Charlotte.

Bud Brinkley

The Gardener's Chairside Reader

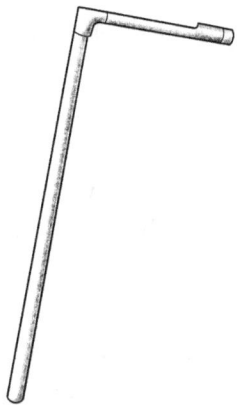

A Super-Duper Garden Seeder

I have never successfully found a seed planter that I liked. They seem to be either over-engineered and over-priced, or they are flimsy in construction and not worth the effort. To be specific, I am talking about the walk-behind push style of seed planters that generally consist of a frame and wheels and various chains, belts, and gears that drive other parts designed to drop seeds at a predetermined depth and distance from each other. Admittedly, part of my dislike for seed planters stems from my own ignorance in how to properly use them as they all seem to need constant adjusting to work properly. The extremely sandy soil in my garden also creates difficulties maintaining traction in order to get the drive wheel to turn while avoiding compaction of the seed bed.

While looking for other solutions and options for planting seeds, I saw a photo in a book of a fellow who used a rubber tube held in his hands and he manually dropped the seeds into the tube. The seeds trickled down inside the tube and fell out at the bottom into the dirt. I liked that idea as it eliminated the need to stoop over and drop each seed into the ground which would save my old back. Of course I knew I could engineer the thing to better suit my needs and so here is my super-duper version...

Bud Brinkley

I used 3/4" PVC pipe for the construction since it is stiff, light weight, and cheap. I based my design on a standard walking stick or cane that you hold in your hands. By slicing off the top section of the cane you can then store a small supply of seeds inside, and easily push them into the drop pipe using your thumb. One hand is all that is needed to operate this planter and you just simply walk along and push the seeds down the pipe and they fall down into the furrow neatly, with almost no seed waste.

The list of parts needed to build your seed planter are as follows;
3ft. long section of 3/4" PVC pipe
6" long section of 3/4" PVC pipe
90deg. PVC elbow
PVC cap
PVC glue

You can easily cut the PVC pipe using an ordinary hacksaw. First start by measuring the length of your drop pipe. Stand upright and measure a comfortable distance from your hand to the floor when your arm is bent in a horizontal position. (Much like measuring for a walking cane.) Once you have this measurement, cut the long section of pipe to length. Also cut a 6" long section of the pipe as this will become your

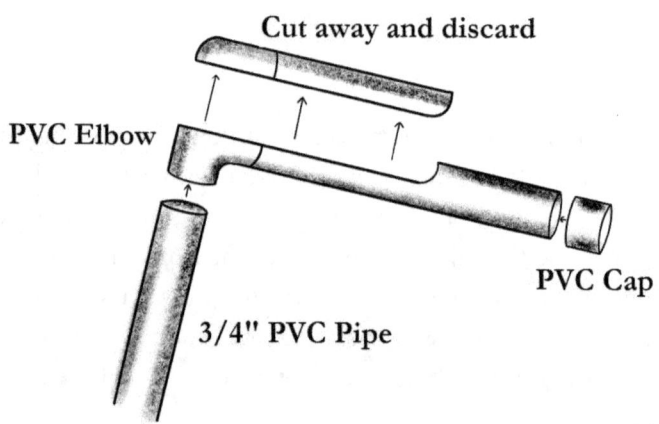

The Gardener's Chairside Reader

handle in a later step.

Next, glue the long piece of pipe to the short section using a 90 deg. elbow. The final step is to glue the cap piece onto the end of the short section of pipe. Let the glue dry for a few minutes and then using a pencil, mark a line for the cutout section along the top of the handle as per the illustration. Cut out this section using a hacksaw and file the cut edges smooth with some sandpaper till there are no burs or snags along the cuts.

Your super-duper custom seed planter is now ready for use! All you have to do is pour a small supply of seeds into the handle section and the using your thumb, gently nudge a single seed up to the top of the drop pipe and watch it fall to the floor inside the pipe. I suggest practice using it on a hard surface floor at first before trying it outside in the garden. Start by walking along and dropping the seeds, and with a little practice, it's easy to get your spacing and control just right.

Now you can go outside and plant even the smallest of seeds with confidence and your back will thank you too!

Bud Brinkley

The Gardener's Chairside Reader

My Failure With Drip Irrigation

I have always had a love/hate relationship with water hose sprinklers. On one hand, it is cheap and easy to set up and you can move it around the garden where it's needed. On the other, it is a very inefficient tool to use for watering a garden. Much of the water output of a sprinkler is lost to evaporation especially on a very hot day. There are other methods available these days to water your plants and drip irrigation is currently the latest fad it seems.

    I bought into the drip irrigation method after watching numerous YouTube videos of folks extolling it's many virtues. It seemed like a great idea and I could easily see how it would save on water consumption by virtually eliminating the evaporation problems that plague the old fashioned garden sprinklers. One downside to drip irrigation is the cost of the equipment and supplies. The drip tape itself is rather pricey but the cost of the many little plastic fittings involved is highway robbery for sure! I have a real problem with paying $5 for a little plastic valve especially when you need 25 of them and they probably cost just pennies each to manufacture. But I won't bore you with my tightwad tendencies for now.

Bud Brinkley

After Mr. UPS Man delivered a big box to my gate, I got busy installing my shiny new irrigation system in my by now thirsty garden. My first problem appeared evident when I tried to unroll the large and heavy spool of drip tape down each row. I devised a method of using a broom stick handle as an axle of sorts and eventually got that sorted out. My next obstacle was connecting the myriad of tubes, tees, and all those $5 valves together. About 4 years ago I had a complete knee replacement surgery and it has been well worth it except for one major complaint - I cannot get down on my knees anymore. This presents a major problem for a chubby old guy like me as I stood there looking at all of this stuff spread out in the dirt at my feet. How am I gonna get this done? I never did find an easy way to get the parts put together, but somehow I did manage to finish it complete with blowing my poor back out for several days while doing it, too.

There are two schools of thought about installing drip tape in a garden. One is to bury it in a shallow trench and plant over it. The other is to leave it on the surface as close to the plant base as possible. The main advantage of going to all the trouble and effort to bury the drip tape is nearly zero evaporation and the water gets right to the plant roots where it's needed. I of course opted for the more difficult method of burying the tape as hey, that's apparently how I typically operate according to my dear wife. I started digging shallow trenches down each row with wild abandon and dancing around while laying in the drip tape and then covering everything over with the loose soil. It became evident that when you have 25 rows in your garden it quickly becomes a lot more work than was indicated by those YouTube extollers. For those readers who are familiar with using drip irrigation, yes-I know there are fancy tractor implements and other tools that can dig the trench, lay down the tape, and cover it back up quickly in a single pass down the row. But these fancy tools are very expensive and and I'm still wondering how I am going to pay for all these little $5 plastic valves. So I did it the hard way primarily to please my wife so she could say, "I told you so!" I eventually got everything assembled and turned on the water supply to the drip irrigation system for the first time. After much hissing, spitting, and sputtering water magically started dripping and oozing out the drip tapes and after a few minutes each row had a very neat wet strip down the center. Hooray!

    Enter the moles, voles, gophers, and the neighbor's mongrel dogs. The problem with putting the water neatly down each row just below the surface is that it seemingly attracts these varmints like a magnet, especially when it is in the dry season of the summer. The little tunnel-digging demons can smell that wonderful cool water from a very long distance and they will make a b-line straight for your precious little seedlings and their tender roots. But they don't stop there! They also start chewing on the buried drip tape and in the course of one night you will suddenly have 15 little geysers of water spewing out of the cuts all over the place. Again, I painfully get on my knees and repair each damaged spot with a special splicer fitting, that by the way cost a mere $2 each. I eventually had so many tears and cuts in the lines that I gave up and started using the sprinklers again. I have not given up entirely though! I am currently experimenting with leaving the drip tape on the surface of the rows to hopefully deter the tunnel varmints. If you don't have problems with these pests in your area, then I could see where the drip irrigation method would be the cat's meow. Around here it just makes my cats dig up the garden looking for a tunnel-digging varmint snack.

Bud Brinkley

The Gardener's Chairside Reader

Morning Coffee and Blessings

If you have a chance, try to slip outside to your garden just after daylight. The quiet mist in the distance and the absolute calm is about to disappear before your very eyes. As soon as the sun rises just a bit, the world comes alive! Birds start their daily singing and foraging, and bees start finding the freshest and choicest flowers in the garden to gather their pollen. The plants usually open up their blooms early in the mornings so that the bees and other insects have easy access to their bounty. Most vegetable blooms only last a few hours throughout the day so the bees and other pollinators get to work early. A vegetable garden can become quite a busy place in the mornings with much industrious work starting-like a busy food factory on a typical work day.

As I sit on the front porch of my little cabin next to the garden and have a cup of coffee, my mind starts running through the day's tasks ahead. As I consider my daily chores such as watering, fertilizing, or weeding, I realize I am exactly like the birds and the bees. I have a job to do just like them. I find it amusing that although I consider even the most menial tasks of "normal" yard work a dreadful chore to be avoided as long as possible, I will joyfully water, cultivate, and pull weeds

Bud Brinkley

all morning long in garden without giving it a second thought. I do love a clean and neat yard, but I take pride in a clean and neat garden.

I am always amazed at how most vegetables seem to grow the fastest at night. I look at my tomatoes, okra, and cucumbers and what was too small to pick yesterday evening is now approaching being too large this morning! It is almost as if God sends worker angels at night to quietly and secretly stretch and tug at the vegetables. I am sure that there is a botanical reasoning for all of this, but I prefer to just sit here and drink my coffee and look out upon my garden and see first hand God's miracles. After I cut and harvest a few large pods of okra, I sit down in my chair on the porch again and take a moment to inspect each pod for bugs, ants and such. I marvel at how quickly it has grown in just one night. Yesterday this okra pod was around two inches long and not big enough to harvest. Now just 10 hours later it is nearly 6 inches long! I take a moment to ponder how something can be created from nothing by using just energy stored in a plant. Think about this. I am holding in my palm, 4 inches of brand new nutritious and healthy plant material that was not in existence yesterday. Cell walls were created and cellular tissue formed overnight. As each cell was formed it was joined to countless others to create a specific fruit that grew and formed itself into 4 inches of new okra pod that will eventually nourish my body and give me yet even more energy in return.

Another morning observation reveals that what seemed like a great idea yesterday looks like a total flop today. As I look at my tomato trellis design, I realize that I was wrong about my great idea to use a hemp string that is biodegradable. This morning the string has already stretched and started sagging so I can see that I will need to replace it with a sturdier version.

I am also questioning my idea about using my new fangled drip irrigation timer that can be controlled by my cell phone. After losing the Internet connection and messing around for probably 15 minutes trying to get everything rebooted and running again, it occurs to me that I could just simply reach down to the water spigot and turn on the garden

The Gardener's Chairside Reader

water manually. Instead of so called convenience, sometimes simple is just better. I chuckle at my foolish and misguided ideas at times. I decide that I will keep using the Internet timer for now though if for no other reason than giving me something to mess with on the phone while I am sitting somewhere else bored out of my mind.

Yes, a morning in the garden is a busy place in both actions and thought. As I walk back towards my house I think about all of the activity and energy being expended to create even more energy and activity. I look down at the little plastic bag full of okra pods that I am carrying and smile at the thought that I too am a worker bee of sorts. Mowing the yard can wait another day or three. I really must get that tomato trellis fixed before it gets too hot and get that Internet irrigation timer sorted out too.

Bud Brinkley

The Gardener's Chairside Reader

A Twisted Plot

My current garden layout consists of 4 plots that are 20ft x 30ft each. This is about all the usable space that I have available in this area at the moment. These days, with my rapidly approaching old age, I want to keep the garden size manageable for me to work. I first worked out this new garden plot on New Year's day and I tried to keep all four plots accurate in size and shape. As the cold wind whipped around my neck I was busy doing survey work in order to establish the plot boundaries. Why I picked a cold New Year's day to do this is a mystery but I suspect it was mostly out of boredom and my desire to get back to spring gardening as soon as possible. Here's how I set it up.

To measure out my garden plots I used a bit of geometry and a handy tool called a wheel measure. You have probably seen utility company folks walking along pushing a stick with a small wheel attached to it along the ground. I purchased one of these handy tools years ago and I have found it indispensable for measuring everything from fence lines to driveways. I recommend every gardener try to purchase or borrow one. These little devices are very accurate and best of all they allow one man operation compared to a long floppy tape measure. The way I ensured that my plots were evenly spaced and shaped like a perfect rectangle is to measure diagonally across the plot to each set of corners. When you get each diagonals exactly the same length, your plot shape is perfect and not skewed or or twisted. An important note to consider as you plan your garden plot size is fertilizer. It is much easier to figure fertilizer application rates when the total sq. footage of your plot is a nice round number. For instance, it is much easier to calculate an application rate in your head for 600sq. ft. than say 465 sq. ft. or some such odd number. This will save you years of aggravation down the road so it's best to work that out now as you lay out your garden plot.

To mark the plot, I made some survey stakes and hammered them in the ground at the appropriate spots. This may seem like overkill, but this is a great tool to keep your plots from growing or shrinking or even moving out of position as you plow the ground for a number of years down the road. You will always have your original reference of where the plot is no matter how much you disturb the soil. To make the stakes permanent and weatherproof I used 2ft. lengths of steel re-bar. For safety and visibility while plowing and tilling I topped each stake with a small hi-visibility orange re-bar cap that you can get at any building supply outlet.

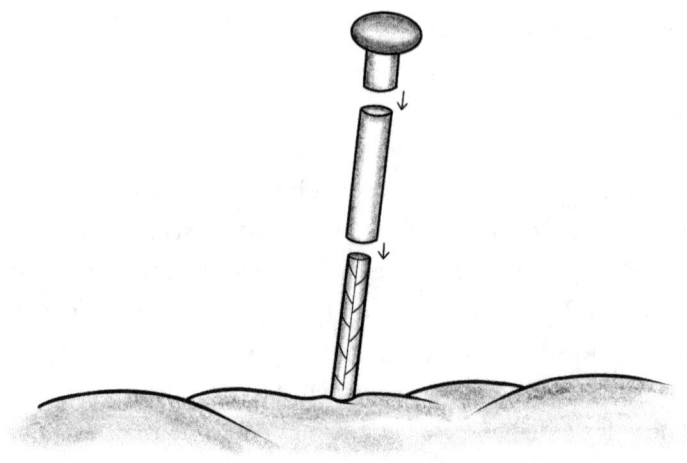

The Gardener's Chairside Reader

One additional bonus tip is to cut a short section of 3/4" PVC pipe about 6 inches long and drop it over the re-bar before installing the cap. This serves a very handy feature. You can drag your water hose around the edges of the garden plot without fear of pulling it over the tender plants and damaging them. The PVC makes the hose slide easily around the stake and it keeps the hose away from the nearby plants.

The are many other factors to consider when laying out a garden such as direction orientation, available sunlight, and others. No garden is going to be perfect all the time but by managing the location as best as you can, your garden plot will be neat and organized. You're welcome!

Bud Brinkley

The Gardener's Chairside Reader

Tomato Temptations

This year I had an amazing tomato crop. I planted three 30ft. rows of Brickyard, Bella Rosa, and Brandywine heirloom tomatoes. I put them in the garden when the seedlings were about a foot tall and strong and healthy. I also added some compost, used drip irrigation, and tried out a new and nifty trellis system. The stars must have all been aligned just right because the plants shot up and grew like crazy.

After a few weeks, I started seeing little green tomatoes grow on the vines. For some reason I have a ridiculous habit of suddenly bursting out singing, "One tomato, two tomato, three…" whenever I see a baby tomato for the first time. Who knows where I started that, but it is just some stupid thing I do when I get excited. Anyway, these baby tomatoes quickly grew and multiplied till suddenly I had probably 300lbs. of green tomatoes on the vines. Wow! Weeks passed as I anxiously waited for the first sign of pink on the green tomatoes. I fussed over them, protected them from the birds and other varmints, and generally worked myself into a froth waiting on this huge crop to mature. I was so obsessed, that I actually had a dream one night about my green tomatoes being stolen

Bud Brinkley

by my father-in-law. While that scenario is not really so far fetched, it did make me wonder if I had gone over the edge a bit.

After a month or two of waiting I decided that I had waited long enough for the tomatoes to ripen. Clearly something was wrong as they were all still very green and hard and showing no signs of ripening or maturity. To make matters even worse, I kept seeing photos on Facebook of folks proudly displaying bushel baskets of beautiful red tomatoes picked fresh from their gardens and some were even beginning to can and preserve their harvest. Yet I still had nothing at all except hard green tomatoes just hanging there on the vines.

One evening I was watching gardening videos on YouTube and I saw a random video of a lady who ripened her green tomatoes in a cardboard box using a banana. Her theory was that tomatoes and many other vegetables ripen quickly when they are exposed to ethylene gas or something like that. Bananas give off this gas when they ripen and if everything is in a sealed box the excess gas from the banana would jump start the tomato ripening process. I had to try this as I was clearly getting desperate.

The next morning I gathered up about 50 lbs. of the green tomatoes and brought them back to the house, carefully placing them in a large cardboard box. I put a well ripened banana in the middle of the tomatoes. I ever so carefully taped up the box so that none of the precious gas elixir would escape and foil my plan. The next day I was impatient and so I peeked inside the box and there was nothing turning red at all as the tomatoes were still green and hard. For the next 4 days I checked on them, quickly losing hope that this plan would work as I could see no pink or red on the tomatoes anywhere. To make matters worse, the banana in the box was getting pretty rotten and the smell was a bit overwhelming.

On the morning of the 5th day I checked on them once again feeling pretty disheartened at this point. I peeked inside the box with a flashlight and I nearly fell backwards. I suddenly had a box full of

The Gardener's Chairside Reader

beautiful red and pink tomatoes! I ripped open the box with lightning speed and agility to see what can only be described as tomato bliss. Beautiful, plump and red tomatoes were in my hands and many others were turning pink and were well on their way to maturity.

It was such a success, I eventually ripened many other tomatoes that way with similar results. About a month later the remaining green tomatoes that were still on the vines finally started ripening naturally. If it had not been for the box method, I would have lost probably half my crop to the birds, squirrels, deer, rabbits, father-in-laws, etc. whom were all eagerly eying my crop at the time. It was a complete success. I don't remember the YouTube lady's name or video title. Whoever you are, you have my gratitude and applause!

Bud Brinkley

The Gardener's Chairside Reader

Gardening From the Comfort Of Your Own Bed! Click Now!

In this day and age, the Internet has become my first source of information for just about anything. YouTube in particular has become my most used site whenever I am looking for garden advice and ideas. YouTube is full of "Garden Guru's" as I like to call them, spewing forth video after video covering all sorts of topics such as starting seedlings in the greenhouse all the way to canning and preserving the harvest.

YouTube in my opinion has created a huge problem of misinformation however good their intentions may be. To elaborate into this further you must look into how all of the Internet Garden Guru's make money from their videos. YouTube channels make money from their videos though advertisements shown usually at the beginning of the video. The idea is that the more views that a particular video gets - the more money is earned. If a channel only gets a few dozen views on a video while a popular channel gets 20 thousand views in one day it is obvious which one will make the most money. This creates an entertainment tabloid atmosphere of being in a popularity contest which rarely has a good outcome. Enter a new phenomenon now known as clickbait. Make your video title into something so garish and outrageous that I have just got to click on it and start watching and you just created clickbait.

I like to relax in my bed in the evenings and watch YouTube videos to try and get tips and ideas for my various hobbies with gardening being one of those. As I scroll through the multitude of gardening videos I see titles such as *"Grow Enough Food in Your Small Flowerbed to Feed a Family of Six!"* or *" He Doubled His Production Overnight Doing Just This ONE Thing!"* My favorite so far has to be the guy who pruned all of the leaves off of his tomato plant except for the last three tiny leaves at the top. I lean over and say to my wife, "This idiot just killed his tomato plant because he was said it needed more airflow or something".

Some channels have literally hundreds of videos. They make videos showing them unboxing a new garden gadget or make a 30 minute video of them sitting in their car yapping about shopping for plants without conveying any actual useful information to the viewer. If they don't publish a new video every few days, their channel popularity quickly fades and the advertising dollars go away. Thus the desperate flooding of worthless and ridiculous videos of them showing how they used old bedsprings as a trellis.

The next issue is experience. Many YouTube gardening stars are obviously in their early twenties at the most, but they consider themselves experts with years of experience. Sorry sweetie, but anyone who does not have the maturity to know that they are pronouncing legumes wrong probably can't teach me anything new. Finally, as we all know, our local climate can have a huge effect in how we grow plants. The twitchy guy in Great Britain preaching that his way is the only sensible method to grow a garden does not have a clue about the enormous insect pressure on our gardens in the deep south of the United States. Yes, I realize that his method works beautifully for him in his short and cool growing season, but he needs to understand that my 100 degree days with 90% humidity will just melt his fleece tunnel

The Gardener's Chairside Reader

coverings. No one particular way works for everyone yet they still expound upon the virtues of hugelkultur compost saving the planet while in my location, it will just feed the termites around here, dude.

Fortunately there are some YouTube channels that actually try to educate rather than just entertain the viewer and you can learn some useful ideas from those. Still it's somewhat of an entertaining train wreck to read the latest video title, " *Fertilize Your Garden Using Your Own Poop and We'll Show You How! Click Now!* "

Bud Brinkley

The Gardener's Chairside Reader

Tractor Commercials

Depending on where you live, you may have seen the TV commercial from one of the big tractor companies touting their latest offerings. This commercial is targeted to those young city couples who just bought 10 acres in the country someplace and think that they may need to purchase a new tractor. In the TV commercial the wife mentions some chores that need to be done, but the husband says nope and goes fishing or golfing for the day instead. So she gets on their shiny new tractor in her perfectly clean work clothes and happily proceeds to mow, plow, and haul hay and feed all day by herself. When he returns in the afternoon she is sitting on the porch with her perfectly manicured fingernails in her still perfectly clean work clothes drinking a glass of cold iced tea. He asks what she has done today and she smiles and says, "Oh, nothing much really." Remember that one? Folks, that's not how it always happens in the country. Here are a couple of stories of real-life country living...

One day I was using my tractor to mow around my property near some big Sweetgum trees. I was driving my tractor under some low-hanging limbs and I would lean forward and squat right next to the steering wheel and drive myself and the tractor under the low limbs and continue mowing. A big limb somehow snagged the collar of my tee shirt, ripped it off of me and proceeded to plow a very neat but painful scratch down my back resulting in a 18" long x 1/4" deep groove right down my spine. In the mirror it looked like I just walked out of back surgery without getting stitched up first. That incident made for some painful sleeping for a long while till I eventually healed. Yes you can mow around trees folks, but don't go under the limbs.

When a man is around 50 years old or so he still thinks he is in his 20's yet his body is just beginning to break down. There's a tendency to brush your new aches and pains off with disregard and that's when the serious injuries begin. This is an example of this happening to me. One day it was excruciatingly hot and humid here in southeast Texas and I needed to hook up my tractor to a discing plow. Now this implement is big and heavy and probably weighs nearly 1000 lbs. The usual way to hook up an implement to a tractor is to back up the tractor to the disc and then crawl down off the tractor and strain and manually pull the tractor and disc together till the locking pins align perfectly. After several attempts at backing up, crawling on and off the tractor and cussing a little more each time you will eventually get everything aligned properly and hooked up. On this particular day I was sweating profusely in the horrible heat and humidity and I was having much difficulty in getting the tractor aligned to the disc. I was already worn out before I had even begun my day's chores. Sweat was pouring into my eyes and stinging like the devil and the mosquitoes were swarming out of the tall grass in the area.

In a vain, last attempt to get this thing hooked up to the tractor, I grabbed a short piece of steel pipe to give me more leverage to manually move the heavy disc into position. I wedged the pipe into position, pulled on it with all of the strength I had left and suddenly I felt a terrible pain and actually heard my back crack. I collapsed and fell down in the tall grass and curled up in a fetal position writhing in horrible pain that I had never experienced before. I was hurting so badly that I started crying and screaming uncontrollably for help, which is bad stuff. I was totally alone out there in the pasture and no one could hear me no matter how loud I screamed. Somehow I eventually managed to get up and hobble back over to my truck parked nearby and crawl inside the cab. I turned on the air conditioning and sat there in the seat doubled over crying in pain and at that point I realized that when I fell to the ground, I must have fallen into a fire ant bed as they were crawling all over my clothes. I think I eventually went into shock from the unrelenting pain and passed out right there in my truck.

A long while later I woke up, the truck was still running, the cab was cold, and the ants were still stinging me everywhere. I somehow managed to drive the mile or so across the pasture back to the house and to the safety of my wife's loving arms. She nursed me back to health, but it took many weeks before I could walk pain free again and I still cringe at the memory of that fateful day.

Yes, the pretty lady in the commercial drinking iced tea after having fun all day on her new tractor is quite the image to enjoy. I imagine city folks watch that commercial and run down to the dealership and buy new tractors all the time. That's not all that bad really, but real life in the country can get you seriously hurt or worse. Please be careful!

Bud Brinkley

The Gardener's Chairside Reader

Water Hose Kinks and Thrills

In my younger days… about 5 years ago, I needed a new water hose for general use around the yard. The scalding hot Texas sun had claimed another victim and the time had come for me to go water hose shopping. My first inclination was to look around online at the offerings because I generally don't like to wander up and down the isles at my local big box store as much as I used to. Sure enough after much intelligent studying and investigating the newest technology, I had settled on a new type of hose that shrinks itself to about two foot in length when not in use, but when the water is turned on it stretches out to around 25 feet or so. You have probably seen this new-fangled hose being advertised on TV during one of those late-night programs. About a week later, a box showed up at my gate and I was excited to try it out.

My first reaction at actually seeing the hose and holding it in my hands is that this thing sure looks cheaply built. But it was light as a feather, took up almost no space, and it came in a really cool fluorescent turquoise color that gave my yard area a tropical beachy feel. I was preoccupied with other things happening at the time so I decided to just keep it and make do with it. One day a few months later, my grass had

Bud Brinkley

grown pretty tall due to my lack of mowing and my old pet goat was taking advantage of this by sneaking around the yard munching on the tasty shrubs-even though she knew they were strictly off limits. I was sitting on my back porch watching her when I came up with the hilarious idea of turning on the sprinkler attached to the hose to scare her away from my shrubs. I casually slipped over to the spigot and reached down and turned on the water full blast.

The hose immediately expanded into a wild turquoise snake that was 25 ft. long and started hissing and the sprinkler head that was attached to the end suddenly flew up in the air like a viper in one of those wicker basket thingys you see in the cartoons. My poor goat, Jezebel, let out a squall of terror and jumped back over the fence and stayed hidden from me for the rest of the afternoon. I was laughing out loud and congratulating myself when I noticed that my hilarious prank on my goat had caused my tropical hose to split and render itself forever useless.

I ended up once again hose shopping at the big box store the next day and finally settled on a "medium duty" hose because the others were way above my tightwad budget. I am not sure how they actually go about classifying a water hose as medium duty or light duty or whatever, but I am sure it is done by an over educated water hose engineer sitting in a cubicle somewhere probably in New York City who had never actually even used a water hose before. As it turned out, it wasn't even close to a medium duty in my opinion and lasted but a few short months.

In a moment of desperation, I remembered that I still had a long water hose at my mother's house on the other side of the ranch. I cleaned it up and was amazed at how well it had held up over the years. It was heavy and thick yet still smooth as a baby's butt and the hot sun had not affected it in any way. I remember buying it over 20 years ago in Sears at the mall. Unfortunately, it has no brand markings and I have no way of purchasing another. I am still using it today because well, it's the best water hose that has ever been made. I keep it carefully coiled and

The Gardener's Chairside Reader

out of the sun because I will never find another one like it. I jealously guard it from my father-in-law's sticky hands and I take joy in using it each evening watering things. Yes sir, a really great water hose really can change your life. Poor old Jezebel is long gone now, but I bet she would have loved this water hose, just like me.

Bud Brinkley

The Gardener's Chairside Reader

Till vs. No-Till vs. My Aching Back

Ah yes, the modern debate of till vs. no-till gardening. This debate between the two opposing camps of farmers is almost as heated as those between the Republicans and Democrats. Each side scoffs at the notions of the others, yet each camp has valid points to consider. My personal view of all this nonsense is to do whatever you want to do in your own garden and leave the preaching in the church house.

Traditionally farmers have tilled their land using various plows and implements and then planted their crops in the freshly tilled soil. Throughout the growing season as weeds began to sprout in the garden they usually till again around their crops to quickly eliminate the unwanted weeds. This sounds simple enough, and it has worked through the eons as man has evolved with more modern machinery to make the task of tilling easier. Of course, we have learned through our history classes in school that this can have tragic and devastating effects on the land as witnessed by the great dust bowl in the Midwest, and the terrible erosion problems in the Tennessee Valley region and elsewhere during the 1920's. Uncontrolled tillage was rampant in these times and the resulting damage is now well-researched and understood.

I'm not sure where the no-till movement started or even when, but I suspect the back to nature movement in the 1960's and 70's gave it a major boost. This method involves no digging or tillage in the garden at all, soil prep and weed control is accomplished by piling on layers of compost and other biodegradable materials to create a wholly natural approach to gardening. This has both advantages and disadvantages mainly depending on your local climate and geographical region where you garden. Lastly, there is a method used mainly by many large commercial wheat and corn farmers that uses machinery to plant the seeds using a minimally invasive technique of seed drills and the like, planting directly on top the decomposing debris from last season's crops. But I will leave that discussion alone as frankly, I have no experience using it.

Let's go back to the tillage camp for a moment. Pretty much everyone these days understands the consequences of uncontrolled tillage and that rows cut into hillsides in the wrong direction without proper soil terracing will cause rapid erosion. The fact is that tillage in a typical home garden does not usually create these problems. It is simply a convenient and familiar method that produces successful results for the average gardener. Various hand tools such as rakes and hoes are easily affordable and available and these make the task of tillage easy for most folks. Where tillage really shines is for those of us who garden in hot or tropical climates with long growing seasons and the corresponding heavy weed and bug pressures accompanying that. Tillage makes it very easy to keep a garden clean and hygienic, as well as keeping bugs and pests at bay with the minimal use of pesticides.

The no-till camp believes that by using layers of natural compost in the garden beds they are creating more fertile soil and the compost smothers out the weeds that do sprout which it surely does. There is no doubt that compost is a rich and natural fertilizer that has the benefit of low cost and eliminating the salt buildup in soils caused by salt-based fertilizers. The lack of tillage does eliminate most erosion problems too

The Gardener's Chairside Reader

and the worms and other sub-surface life will thrive in these conditions which helps the soil even further. This camp also tends to tout that everyone in the world must use this method right now to save the planet and the tillage camp folks are dead wrong anytime they break out the hoe for a morning of weeding in the backyard. This is where things go awry and the friction starts.

The fact is, the majority of the successful no-till folks live in colder climates with shorter growing seasons while the tillage camp crowd usually lives in hot and weedy areas such as the southern and tropical areas of the continents. I can tell you with a full degree of certainty that no-till did not work for my gardens here in Texas as the weeds, termites, bugs, and other pests feasted on my compost laden beds and eventually the whole thing turned into a giant weedy and snake infested mess. I am also confident that if my gardens had been located further north in the country that I probably would have been successful using the no-till system.

Farmers of any type are independent folks by their very nature and they don't particularly like someone telling them how to do things. The no-till crowd in my opinion should accept the fact that their methods of gardening are wonderful in areas where it has a chance of working well, but not everyone can grow vegetables that way and so no ONE way is the right way for everyone and the planet. My aching back has little affinity for shoveling mountains of compost each season and dumping it in the garden and by necessity, using a wheel hoe or a garden rake for a few minutes each day is about the only way I will ever grow anything around here.

Before I stagger and fall off my soap box I want to bring up another point and that is both camps do have one thing in common at the end of the day and that is our love for gardening and producing our own healthy and nutritious food. Let's all unite with that idea and leave the silly arguments to the political arena.

Bud Brinkley

Cover Crops

Let's talk for a bit about cover crops. Making the effort to incorporate cover crops into your garden planning is a good thing. A cover crop is a sacrificial crop that is planted for various purposes. Two main purposes for me at least, are weed suppression and soil enhancement. Cover crops are not normally edible and are not considered a "cash crop". The basic idea behind using a cover crop is to plant between vegetable growing seasons as a way to naturally enrich the soil and also to organically prevent weeds from sprouting by shading the plot and depriving the weed seedlings from sunlight . Depending on where you live, various cover crops may be used, and some common varieties include buckwheat, sorghum sudangrass, winter rye, and clover. These all work well at weed suppression and building organic matter back into the soil.

Peas are another great cover crop and one that has been largely overshadowed by the above varieties these days mainly due to large scale farming shifting away from using peas in favor of other slower growing cover crops. Peas do grow very quickly and that can be an advantage for the home gardener, especially in areas that have a short window of opportunity to grow cover crops between seasons. I prefer peas because they are legumes and the naturally fix nitrogen into the soil as well as adding green manure fertilizer when tilled under. Peas are also very easy to plant and manage for the home gardener. There are many pea varieties available to use, but for cover crops I just use plain black eyed peas as found in the grocery store isle. There is no need to fret about finding peas to plant at the seed houses. Just buy a bag of dried black eyed peas at the grocery store and you are ready.

To prepare to plant your cover crop, you must clean your garden plot of old plants and other debris left over from the previous growing season. Since I don't use raised beds, I just till my plots using a rotary tiller after removing as much plant material as possible. Some vegetable plants can be left to decompose in the soil such as lettuce and cabbage, but many others need to be pulled out of the ground and removed entirely from the plot. You do not want to encourage the possibility of plant diseases and pests to remain in your plot. By removing the plants and roots you are maintaining good garden hygiene.

After tilling your plot, the next step is planting. For peas, I prefer to just broadcast them by hand evenly across the plot, no need for rows. I then lightly rake over the seeds with a thin covering of soil and water the entire plot well. Keep the soil moist while the seeds germinate, usually between three or four days. Once the seeds have sprouted, water the plot several times a week to keep the soil moist. Peas grow exceptionally fast, and in about 4 weeks you should have a healthy and thick stand of green pea plants about a foot tall.

After about four weeks of growth the peas will start to bloom. At this point it's time to start keeping a close eye on the peas and get ready to till them back into the soil. Peas have one major disadvantage as a cover crop, as they will produce pods quickly that will drop peas to the ground and start the reproduction cycle all over again. It is important to till the pea plants into the soil just after they bloom, but before any pea pods form. This will break the cycle of unwanted errant pea plants growing in between your vegetable plants next season.

The Gardener's Chairside Reader

To incorporate the plants into the soil, first I mow down the pea plants as much as possible using a standard push mower. I then immediately follow behind this with a gas rotary tiller to chop up the stems and roots and to bury the green manure under the soil surface. It's very important to note that you must bury the green plant material as quickly as possible after mowing. If you wait days or even hours after mowing, the sun will have robbed most of your nitrogen from the green manure laying on the surface. Get those plants tilled under as quickly as possible!

Peas are a great option for a cover crop that is easy to manage, to cut and to till back into the soil. Some other types of cover crop grasses can grow quite tall and thick which make it difficult for the home gardener to cut and till without specialized equipment (such as flail mowers etc). Peas can be bought in nearly any store, and if you have some left over after planting, throw them into a pot and have them for dinner that evening!

Bud Brinkley

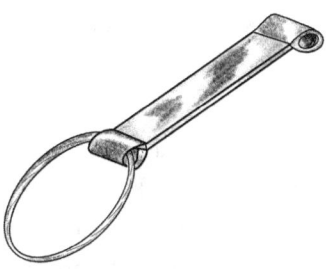

Iron Relics From the Past

While running the cultivator through my garden today I heard a metallic sounding clink come from the ground and I immediately knew what it was. I leaned over and dug around in the loose soil till I found the culprit. In my hand was an old rusty piece of iron from a horse-drawn plow frame that was used here many years ago. I have collected a number of these pieces from my garden plot and I display them sometimes for visitors. Most of the time they lay quietly against the porch post of my garden cabin.

My garden is located on a plot of land that once used to be a homestead from around late 1800 or so. My relatives told me that a big old two story house used to stand on this spot until it burned somewhere around the turn of the century. Once my wife and I were walking in this area and we saw some spider lilies still growing and blooming in a neat row along what was probably an old fence line around the house yard. Somehow those lilly bulbs kept growing and reproducing for over a hundred years in that spot long after all of the structures and people were gone. We treasure the thought of an old homestead being here. Think of the memories of kids playing, growing crops, Easter parties and celebrations, and family tragedies that occurred

Bud Brinkley

in this very spot, now only marked with some flowers growing in a row in the ground.

Today it is hot and humid and I am sweating and working hard in the garden. I stop to sit on my cabin porch and drink some water. I began to inspect the old piece of iron that I dug up earlier. I wash the loose dirt off in the sink and take a closer look at the piece and I can see the faint markings of where a plow horse harness rubbed a shallow groove on the side of the iron brace. I was immediately taken back to the same spot on the same kind of hot and humid day in the late 1800's where some poor old farmer was sweet talking his tired mule to pull just a few rows more. I could see his big old calloused hands expertly guiding his plow through the sandy soil to make perfectly straight rows that are a source of pride and satisfaction for any farmer. I could see the old mule hoping to get back into the cool shade of his barn stall to lean against the wall and take a snooze.

I could smell the scent of roasted chicken in the air that his wife was cooking and she was standing on their front porch shouting for him to stop work and come inside to clean up and eat dinner. I could hear the laughter of little children playing hide and seek around the giant old Sweetgum tree that still proudly stands here today. I could see the family sitting around their table saying grace. I could hear the farmer telling his wife that if they don't find the money for fertilizer soon they will lose the year's crops and they are having problems with the same cutworms that I am battling today. I could see his wife's gentle hand reach under the table and quietly hold and reassure her husband's hand. I just know that all of that and more occurred on this very spot so many years ago yet all that remains are a few flowers planted in a neat row amongst the grassy shade of an ancient old tree. The family's voices are silent now, but their lives will be forever tied to this small patch of land.

It is my hope that 120 years from now after I and my home are long gone, some young newlywed couple will be walking along this same path holding hands and when they see the giant old pink magnolia tree that my wife and I planted from a tiny seedling and say, "Look! This must

The Gardener's Chairside Reader

have been an old home place at one time!" that they will hear my story too.

Sometimes things just speak to me, and that day I listened. I quietly lay the rusty old piece of iron to rest against my cabin porch post with all of the quiet reverence and respect that it deserves.

Bud Brinkley

The Gardener's Chairside Reader

The Feedstore Redlight District

One of my most treasured childhood memories was going with my Dad to the local feed store in our little town to pick up some cotton seed meal cake to be used as fish bait. I vividly remember sitting on a bale of hay in the feed store taking in the aroma of the various sweet feeds dripping with molasses syrup. To make things even better for this 10 year old boy, were all of the old crusty farmers who were gathered around a rickety old table playing dominoes and drinking thick black coffee. I imagined myself equal to their years of experience and stories and I was sure that I could have taught them a thing or two about playing dominoes as I was a fairly good player myself back then. My Dad was back in the feed warehouse digging around for the bag of meal cake amongst the mountains of feed sacks and so I remained in the front of the store taking it all in.

I watched as a skinny old man in overalls pulled a tin of Prince Albert tobacco from his chest pocket and started rolling his own cigarette. He carefully licked the edge of the cigarette paper to activate the sticky goo and using his big shaky hands he started tapping the shreds of dried tobacco from the can onto the paper. After twisting the

Bud Brinkley

ends of the paper closed he had a neatly rolled skinny cigarette and smoothly and casually struck his match against the leather sole of his shoe and went back to playing dominoes again. The fat lady behind the counter kept an eye on me so that I did not wander off into the street as I was sometimes prone to do in those days. She gave me a cookie and a cold soda water probably to keep me still, till my Dad returned. On the way home in the truck I remember my Dad telling me that those old guys were all mostly retired and had nothing better to do with their time except hang out at the feed store. In my young mind they were giants among men. I probably tried to light a kitchen match using the rubber soles of my sneakers too when I returned home.

These days I still have pretty much all of the same feelings and emotions when visiting a feed store. There is of course the same sweet aroma of molasses syrup from the feed sacks in the warehouse. That's the first hook to get you to hang around right there. If you are ever on a diet, for Pete's sake don't visit a feed store as you'll probably end up sneaking a handful of sweet oats from the barrel. The next temptation is the general atmosphere of the place. What's not to love about an ancient old building located right next to the railroad tracks? The well worn wooden floors and old clapboard walls covered with 100 years of cobwebs are something that you just can't buy. The shelves are always stocked with merchandise that looks to be straight out of a 1945 Norman Rockwell painting. There is everything from giant wooden mouse traps to copies of the Farmer's Almanac to thick leather dog collars and real cowbells. This is intoxicating stuff folks. You forget all about your budget as you just have to buy that shiny new grain scoop, and who doesn't need a new -fangled fence stretcher these days?

The old fat lady has usually been replaced by a cute young thing, which can be either a plus or a minus, depending on if she actually knows what cotton seed meal cake is. The old guys sitting at the table playing dominoes are all long gone and replaced by a generation or two

The Gardener's Chairside Reader

later of retirees sitting around talking about their 401K accounts. Still it's a temptation to just find a bale of hay and sit down and take it all in. Feed stores, in my opinion, are the true red light district for old geezers like me.

Bud Brinkley

The Gardener's Chairside Reader

Where Did All Of the Jars Go?

Whenever I think of canning and preserving vegetables, I have this image in my head of some chubby old grandma wearing a flowery apron working next to a wood stove in a 1920's farmhouse kitchen. All of her blue and green tinted jars are sitting on a wooden table covered with a bright red checkered oil cloth. The old zinc jar lids have been cleaned and are sitting next to a bowl of beautiful fresh-picked green beans. Grandpa is visible through the open kitchen window far out in the fields plowing on an old two-cylinder John Deer tractor while a wisp of smoke drifts from his corncob pipe. On the open windowsill is a fresh apple pie cooling down and the old hound dog is lazily sleeping at her feet under the table. Yep, that's what I think of each time I start canning vegetables in my kitchen.

This year my nostalgic daydreams have been rudely interrupted by the Covid-19 virus pandemic. I am not happily inventorying my jars, lids, and rings and getting vegetables washed and prepped. Usually in times past, we have always had plenty of jars and rings as they are re-usable each year. I would always make a leisurely trip to the grocery store to get a few boxes of new lids which only have one use each

Bud Brinkley

season. Because of the food shortages caused by the pandemic, this year's canning took on a much different if not desperate tone. I knew that I did not have enough jars because we were planning to preserve much more food than in years past. Who knows what this winter will bring in food shortages, so I wanted to have a well stocked pantry and be prepared for whatever may happen. I had also started hearing conversations that canning supplies were getting in short supply as it seems many other folks had the same plans as I did. My first proof of this happening here locally was during my weekly trip into town.

My first stop was at my favorite grocery store and as I walked up to the spot where the canning supplies are normally displayed, the shelves were already completely empty save for a few boxes of tiny jelly jars that don't count anyway. A hastily written sign was taped to the shelf apologizing for the shortage. Hmmm, I am beginning to believe what I have been hearing for the past few days. I drove around to another big grocery store outfit in town and the same thing was happening there. Yikes what now? After I returned home, I immediately turned on the ole computer thinking that surely I can just order all my supplies from one of the Internet retail giants because they surely must have thousands of cases of jars stored away in a warehouse somewhere. Nope. Everything had already been sold out. No jars, rings, or lids were left. To add even more angst, other odd items such as vacuum sealer attachments and vacuum bags were either completely gone or the last few that were available were priced so ridiculously high by unscrupulous price gougers that I knew that we were in some serious trouble this year.

I started making jar scouting trips to my local stores in town every few days and occasionally I would score a single case of jars or a small box of lids. One day, I was striking out at every store in town as they all had empty shelves and they all had that same little apologetic sign posted. There was one last small grocery store in town to check before giving up entirely and so I headed over there in haste. I walked up to the canning supplies area and I could not believe my luck. Right there in front of me were fully stocked shelves of beautiful jars and lids in all of their glory! Apparently the store had just received a new shipment and

The Gardener's Chairside Reader

my timing was just right. I ran back to the front of the store to get a shopping buggy with lightning speed. I nearly crashed the buggy into the beer display and came back screeching to a halt at the canning isle and shuddered in disbelief. Standing there in front of the holy grail of canning supplies were several of those chubby old grandma's in their flowery aprons about to fight each other over my loot. They were eying each other with contempt and each one had full intentions of getting those jars in her basket first. Their skinny old farmer husbands were probably waiting outside in the truck smoking their corncob pipes completely unaware of the drama unfolding inside.

Jars or no jars, I was raised to be polite, so I just held back a bit. I let those ladies work it out amongst themselves which they eventually did. Yes, fortunately there were a few cases of pint and quart jars left so I snatched those up and headed to the checkout counter leaving behind store shelves that were bare once again. I felt kinda guilty for buying the last cases, so I was glad that I had been wearing my Corona Virus safety mask so no one recognized me and I slithered home like a snake in the grass. You can bet that I will be searching for canning jars at garage sales throughout the coming year as the last thing I want to see are those chubby old grandma's slinging off their flowery aprons and giving each other a whoopin' in the grocery isle.

Bud Brinkley

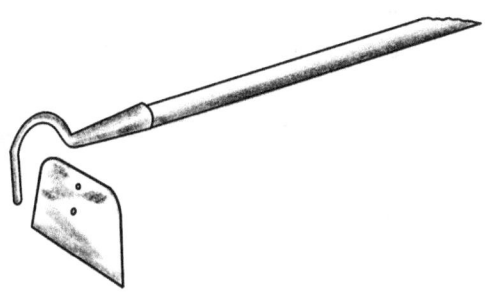

Hoe-Down!

Commonly in the garden, you find the need to weed in between the individual plants. This is a task that can become somewhat tedious if the plants are spaced closely together. There are various commercial made hand weeding implements, but for those of you who like to recycle or re-purpose and save some money, there is a way to make a dandy weeding tool using an old broken garden hoe. I have always seemed to hang on to old or broken garden tools long after their useful life has expired. I am always telling myself that the broken tool might become useful someday, and by golly this is the day!

The idea behind this implement is that by using a small single tine you have much better control at removing small weeds that sprout in hard-to-reach areas. Going one step further by sharpening a razor edge on each side of the tine, it's rather easy to slice the weeds off at ground level. To make this tool you will need an old discarded hand garden hoe. These hoes are generally constructed of a single spline and a blade that is either spot welded or riveted to the spline. Use whatever means necessary to remove the blade until you are left with a single spline. Using a file or a 4" grinder, sharpen each side of the spline to a razor edge. I suggest you then give the whole thing a quick spray of paint for protection.

Bud Brinkley

To use your new tool, just carefully place the sharp spline in between the plant stems and dig out the weeds, being careful to avoid nicking the stems of your plants. The sharp edges make it easy to saw stubborn weeds off at ground level. With some practice this will become one of your most often used weeding tools because it takes such little effort and is so effective. Using this tool just a few minutes each week will go a long way to keeping your garden clean and weed free and best of all it was free to make!

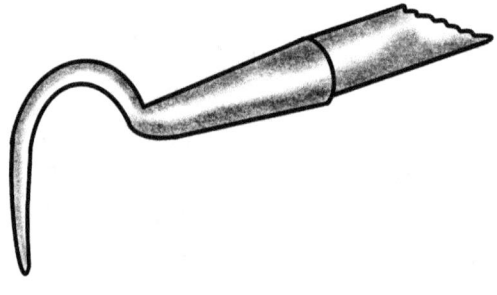

The Gardener's Chairside Reader

Crooked Rows, Moses, and Pearl Beer

Since the time when Moses first learned to scratch the ground with a crude stick and plant something, he probably had a nosy neighbor walk up and tell him that his rows are crooked. I sometimes ponder if that is when he decided to part the seas and move to a quieter spot. Crooked rows are the embarrassment of any farmer and yet still today I see crooked rows everywhere. Not my garden, but nearly everywhere else. One thing I have learned about "drawing rows" as the old timers used to say, is that you need to draw your rows on a day that you wake up happy, sober, and clear headed. This is a real art form that is to be perfected and passed along from one generation to the next. My poor old Pops tried everything he knew to keep his garden rows straight, but the Pearl Beer he drank throughout the day usually caused his rows to be "crookeder than a dog's hind leg".

I remember one hot summer day on our farm my Dad and I were going about the business of starting a new garden plot. If I remember correctly this garden was closer to the house than normal for some reason and was around a couple of acres in size. We were comfortable when I was growing up, as my Dad worked very hard to provide us with

Bud Brinkley

a solid middle-class upbringing and so we did have several luxuries such as one of the very early riding tractor style lawn mowers. It was an old Sears lawn tractor that was very loud, vibrated madly, and was very difficult to drive. Back then, the little lawn tractors had a stick shift and a clutch just like a car and you had to be really careful or the thing would jump in gear by itself and run over you. Our old Ford garden tractor was broke down in the barn as usual, so my Dad and I concocted a scheme to use our little Sears lawn tractor hooked up to an old horse-drawn plow to draw our rows for our new garden. I was around 10 or 12 years old at the time so I didn't know any better.

We got the plow hooked up to the lawn tractor and our plan was for me to drive the tractor while Dad walked behind the plow attached to the lawn tractor by a short chain and he would guide the plow. In a sense, we were using the lawn tractor as a mechanical plow horse of sorts. I distinctly remember him signaling for me to engage the tractor clutch and start moving forward as it was so loud on that little beast of a machine that I could not hear him well. I hated driving that thing as my legs were too short to reach the clutch and brake pedals without sitting right on the edge of the seat. I managed to snap the clutch pretty roughly and we lurched forward. We ended up making the first few rows rather nicely in spite of my sloppy driving. Dad was probably feeling pretty confident by now and probably congratulating himself on his neat idea (as I am so prone to do myself these days). My Dad was steadily drinking that cold Pearl Beer all afternoon though and by the end of the day his rows got more and more crooked. I mean bad crooked too.

Since our new garden plot was now right behind the kitchen window my Mom had a good view of everything that we were doing outside that day. That evening as I was outside playing after supper, my Mom told Dad that the rows were too crooked. Dad and Mom kinda got into an argument over this and so the next day Dad told me that we would have to re-plow the garden. The plan was to draw straighter rows to please Mama…rows that weren't so "angagogglin" as she used to say. Again my short little legs accidentally popped the clutch on the tractor

The Gardener's Chairside Reader

but this time it took off speeding down the garden dragging both the plow and my poor Dad flailing around in the dirt behind it. He started yelling at me, and at this point I would had rather been anywhere except right there at that moment. I remember no matter how hard we tried, our rows looked terrible and misshaped. I was hot and dirty and probably crying, Dad was fairly drunk, and we both finally gave up by the afternoon.

The garden stood in silent shame for a few days. After Dad left for work one morning, Mom asked me about our problems in the garden and I told her I hated that stupid tractor because I could not reach the pedals, and it was noisy and it scared me. She patiently talked with me and after seeing my problem, she came up with the idea to tie a block of wood onto the pedals so I could reach them better. This method worked well for me back when I was learning to ride a bicycle so it seemed like it would work good on this tractor, too. After getting the lawn tractor pedals rigged up with the wood blocks, Mom and I tried it out and she taught me to drive it around the yard much better and smoother. She had the patience of a saint with me.

When Dad got home from work we were all set to try drawing rows again. After the plow was hooked up to the little tractor Dad probably rather nervously gave the hand signal and I turned back around and nervously gripped the steering wheel and eased the tractor forward slowly and surely this time. I do remember that we re-plowed the whole garden plot that evening. Meanwhile, Mom was busy cooking a big supper and was probably watching my every move through the kitchen window.

Of course I am sure that as typical red-blooded men, Dad and I took all the credit for our success in the garden that evening. I bet we sat around the table eating fried chicken and mashed potatoes and talked about fishing and hunting and other manly pursuits. Mom probably secretly slipped outside while we we yapping and popped open a can of Dad's cold Pearl Beer and and walked through her garden to make dang sure that her rows were no longer "angagogglin."

Bud Brinkley

The Gardener's Chairside Reader

The Wormville Motel

My dangerous habit of watching YouTube videos put a new idea in my head that I should start raising worms. Yes, you read that correctly. The YouTube gurus taught me that by raising a batch of worms in a container, I would soon have mountains of wonderful rich worm castings (worm poop) to make my plants grow to unimagined sizes and I would become a better Earth citizen in the process. I reasoned that the methods they employed would be rather easy to accomplish myself and that in no time I would also have those enticing piles of worm castings to spread throughout my garden. It all seemed rather simple, if not somewhat primitive, and it looked like a fun project, so I commenced to building my own wormville motel.

As I began researching, I noticed that many of the YouTube videos on raising worms were produced by folks in some the smaller tropical countries and that they tended to raise worms on a much larger scale than my desires. It was obvious that these folks were selling the worms and their castings commercially. Since I only wanted to start on a smaller and more personal level, I sought out the more obscure videos of people who were raising worms for use in their personal gardens.

Bud Brinkley

It became evident right away that the small scale guys had wildly odd ideas on how to raise worms. I saw everything from feeding their worms with old newspapers to rotting carpet soaked in beer, so I had to weed through the videos produced by the amateurs and mostly figure out on my own what would work. The commercial worm farm folks seemed to have their methods finely tuned for success so I tried to adapt what they were doing to my own meager ambitions.

The basic premise of this whole operation is that you purchase a bag of worms from your friendly worm farmer and place them inside a container or tub and then feed them. The little buggers then eat their way throughout the food and they poop. A lot. You then harvest their worm poop and spread it as a rich natural fertilizer on your plants. The worms continue to eat, poop, and reproduce and the process continues on. This sounded simple enough. Instead of opting for the usual plastic storage tub that most of the amateur folks were using, I of course got a bit carried away. I spent a full week in my woodshop building a fancy three level worm hotel out of cedar wood and I even stained it, added some decorative trim, and made it look rather pretty almost like a piece of fine furniture. On my workbench stood what I thought was the nicest worm motel around and my worms would be very appreciative of living in the Ritz Carlton and thus, would produce their castings in abundance.

The next task at hand was to find some worms. To my surprise, I found many vendors selling worms on eBay so I ordered my first bag of worms. They showed up in my mailbox three days later neatly packaged in an inconspicuous little cardboard shipping box. I couldn't help but wonder what my mail lady would have thought if she knew she was carrying worms in the seat next to her in her little car. Anyway, I carefully inspected my worms and they all seemed healthy and alive if not a little sluggish from their long journey across the country.

The worms settled in their new home nicely and I started feeding them vegetable scraps. This worked only marginally well because much

The Gardener's Chairside Reader

of the scraps were left uneaten and they started rotting. Searching for a solution, I watched yet another video of a fella who ground up his vegetable scraps in a food processor first to help the worms with their little tiny mouths eat their food better, so I gave that a try too. This did actually work better so now it is commonplace to see me in the kitchen grinding up kitchen scraps into a yummy greenish worm milkshake. I have had my worms for about 5 months now and I am still going through my daily routine of feeding them the green milkshakes and occasionally a rotten banana (as that's like worm crack). They go absolutely nuts over rotten bananas.

I have yet to successfully acquire enough worm castings to use even in a small flower pot, but by gosh I do have the happiest worms in town and I just know the wormville motel will be in full production someday. Little tiny baby worms are being made and I have singlehandedly created a worm nirvana that I just know that will someday become an empire of poop production.

Bud Brinkley

Life In Three Days

Plant seeds are an incredible miracle when you think about it. Stored within this tiny little package is everything needed except water to produce life. By adding moisture around the seed something inside wakes up from a deep slumber. The hard and dried seed shell begins to swell up and soften just a bit and soon a radicle or future root begins to form. This tiny little finger instinctively begins to reach downward into the soil that is surrounding it to anchor the seed pod and give it some stability. The seed pod then usually splits and the next thing you see is a tiny plant seedling emerging from the soil just starting it's young life. The first leaves are usually called false leaves but they are really cotyledons that show up first before the plant forms it's first set of true leaves.

Of course, the details of seed germination are quite a bit more detailed and complex than my explanation, and I am still learning more about the process as I can. I always enjoy planting peas and seeing the little cracks in the crusty earth on the second day with the pea seedling suddenly rising from this crack on about the third day. As a kid in school, I always looked forward to sprouting seeds in science class. We would put a bean in a folded and moist paper towel so that we could watch and study first hand the process of seed germination.

Up until several years ago, I took seeds pretty much for granted. Typically, each year you can purchase a packet of seeds from your local garden center and you are able to grow vegetables reliably each season. This year was quite a bit different because of the Covid-19 pandemic. Everyone (and their neighbors, too) decided to try to grow a garden because of the sudden interruption in food supplies. Seeds became in short supply and downright difficult to find. Now I look at seeds in a different light and realize that someday sooner than I ever thought, the little seed in my hand today may be the difference between starving to death or growing my own reliable food source for tomorrow. Yes, I know that sounds pretty severe but in the not so distant past people did starve to death because they had no seeds. It's that basic and that important. It is so important in fact, that research foundations have created actual underground seed vaults to safely store millions of varieties of plant seeds in case we ever experience a catastrophic event like nuclear war or worldwide famine. Those seeds that are stored in carefully controlled and cool environments will be our only surviving link to be able to restart our worldwide agricultural food supply.

This year because of the sudden seed shortage, I decided to start saving my own seeds from the excess vegetables that I grew. I harvest the seeds from the fruit and allow them to dry. I then carefully store them in paper envelopes to avoid mildew and label them. Now I don't have an underground seed vault or anything fancy. I just keep them stored in a dark and cool place ready for when I need them. Unfortunately some hybridized vegetables won't produce reliable fertile seeds so I have taken an interest in studying some of the older heirloom varieties of vegetables. I prefer to save those seeds instead of the more modern hybrid plants that are more commonly found these days.

Seed science is a fascinating and humbling study and I highly recommend it to any aspiring gardener. There are websites and books that describe in detail the process of seed germination and some outline

The Gardener's Chairside Reader

the methods used to reliably and safely store your own seeds. Take some time to store your own seeds for peace of mind and for your family. And while you're at it, take the time with your kids to put a dried black-eyed pea from your pantry into a moistened paper towel and have fun watching life spring forth in just three short days.

Bud Brinkley

The Gardener's Chairside Reader

Green Acres

Some of you older folks probably remember the funny TV show from the 60's called "Green Acres." My wife and I have recently discovered it again after searching for shows online because most modern TV shows these days either make us angry or disgusted. The basic theme of the show is about a New York lawyer and his wife who move to an imaginary country town called Hooterville to become farmers. They have a hard time adjusting to the sometimes quirky country neighbors. I remember watching this show on TV as a kid and laughing at all of the funny scenes. This show was a spin off of the popular series called "Petticoat Junction", and I loved that show also. Primarily because I loved the old steam engine train called the Hooterville Cannonball. Both shows are still fun to watch, but as an adult, I can see some real comparisons between Green Acres and where I live now.

Whoever came up with the idea for Green Acres and wrote the script must have had some actual real-life experience living in a rural area. The scenes illustrating the difficulties of getting a simple telephone installed or the one about having no electricity yet and having to use an

Bud Brinkley

ancient old generator and a thousand glowing hot extension cords in the house certainly strike a familiar chord with me. Every time I watch Eddie Albert climb to the top of the telephone pole to answer the telephone I cannot help but belly laugh and slap my knee. While my phone is fortunately installed inside my house, I certainly remember the difficulties I had getting all of the original utilities turned on when we moved to this property about 20 years ago.

My troubles started when we got our property surveyed and we needed to file this new survey paperwork with the county at our local courthouse. I dutifully went to town and met with the county fella who was officially in charge of such stuff. His assistant working at the office service window was a real peach. She managed to screw up the paper work repeatedly for several weeks. As I recall, she nearly ended up legally registering our entire property in some stranger's name and it took many frustrating trips to their office eventually ending with a "throw down hissy-fit" expertly performed by my sweet wife before they got it all straightened out.

Next up on the list was getting a septic system permit. I headed back to town and yep, it was the same guy and the same peach assistant at the customer service window again only this time they were wearing different hats and working in the neighboring department at the courthouse. Again, it turned into a giant fiasco of he said vs. she said between the two of them. I finally got my precious permit that allowed me to spend thousands of my dollars on a new septic system devised by them that was way over-engineered and still plagues me to this day.

After the septic was finally installed, it was time for me to get the electricity service hooked up. Our local electric company sent a pleasant young man out to inspect our property and start the paperwork process. All was going well that day until I told him I wanted to have the electric utility lines buried in the ground instead of strung along unsightly poles as normally done. He shuffled his feet in the dirt for a moment and looked at me and said, "Do you know what that will cost?" I grudgingly relented and we agreed to run the new electric service lines on poles just

The Gardener's Chairside Reader

like everyone else.

A little later while walking around my property, this young man spotted an old rusty pickup truck that I had sitting abandoned off in the bushes behind my barn. He jumped with excitement telling me that his hobby was restoring one exactly like it in his spare time, or something to that nature. We ended up making a deal right then and there and I sold him that old rusty truck real cheap that day because he needed it for spare parts and, frankly, I didn't need it at all. A few weeks later, the electric crews showed up to install the new lines. I was completely surprised that they brought with them a giant trenching machine and yes, they were going to bury my electric lines! I'm still grateful to that nice young man for that huge favor.

My wife and I still watch "Green Acres" on TV each evening as we eat our supper and I still laugh when Mr. Haney tries to hoodwink Mr. Douglas into paying for something he needs. Sometimes when I drive around our property and I see the survey markers in the ground I start humming, "Green Acres is the place to be… Faaarm livin' is the life for me…."

Bud Brinkley

The Gardener's Chairside Reader

A Day In The Sunshine

I make it a practice to pray and give thanks to the Lord for each day on earth that he gives me. Everyday, even the bad ones, are a blessing and I try not to waste a single day. Yet some days are just so gloriously perfect. I recently experienced such a day. I live a fortunate life and I know it. Here is my proof...

It was a hot Summer morning as I mowed around my garden area and swept the porch of my new little garden cabin. Everything looked clean and pristine and well, just perfect. I walked back up to the house and told my wife that it would be a great day to leash up the pups, grab a cold watermelon out of the refrigerator and walk back down to the garden cabin. Our dogs know that whenever we put a leash on them that they are in for a special treat as they absolutely love to walk to our garden area. They pulled like pack mules trying to rush down the walking path with their eyes open wide with wild excitement.

We managed to get the pups settled on the shady front porch of the cabin and I cut open the watermelon spilling the sweet juice everywhere. I sat there laughing and talking with my beautiful wife about the goings

Bud Brinkley

on in the lush garden in front of us. I suddenly had one of those "frozen-in-time" moments. As I sat there looking at her fuss with the dog's tangled leash, I watched her face glow with happiness. I looked out at my morning's yard work and admired my little Shangri-La that God had blessed me with. The vegetables in our garden at our feet were just beginning to produce abundantly and as any gardener knows, that's when the plants look their absolute best. I felt healthy, very happy, and very alive. At that moment I was the richest man in the world and I just sat there taking it all in.

We eventually finished eating our watermelon and the pups were getting anxious to get back home. They were pulling on their leashes and dragging their poor Mama with them ahead of me while I was lagging behind carrying my old dog in my arms. I stopped, and looked back at my beautiful garden and cabin and smiled. I turned around and looked ahead at my silly pups with my wife chasing after them. I looked down at my old girl in my arms with her loving eyes, and at that very moment in time I said a quiet prayer of thanks. Life is wonderful.

It's a Tilthy Business

Tilth. It's a funny sounding word that's not used much at all in our language. What does it mean and why should you use it in your daily conversation? To a gardener it's the difference between a bounty crop or malnourished withering plants. To a non gardener person it sounds like you are talking with a speech impediment. I jealously guard my soil tilth like a fanatic and try my hardest to keep it the way nature intended.

Soil tilth by definition is the physical condition of soil, especially in relation to its suitability for planting or growing a crop. Factors that determine tilth include the formation and stability of aggregated soil particles, moisture content, degree of aeration, soil biology, rate of water infiltration, and drainage. I copied that definition from the Internet and it's a mouthful for sure. In simpler terms, soil tilth is the physical health of your soil that allows it to sustain plant growth. This one factor alone can be negatively influenced and easily thrown out of whack if you don't understand what you are doing to your soil. I'll give you a typical example of destroying your tilth.

Bud Brinkley

Let's say you just bought a shiny new garden tiller and you are excited to try it out in your brand new garden plot. This ground has been laying fallow and unused for just about forever and it's covered in grass and weeds. Naturally, you want to use your new tiller to break up and loosen the soil and make the weeds go away. Five gallons of gas later, your tiller has chopped up the weeds and now your soil is nice and fluffy and has the consistency of sugar. In go the seeds, and after a few days and much watering you are rewarded with a few seeds that have sprouted but your germination rate was not great, as many seeds never broke the surface and are just laying there. That's your first example of bad tilth. The soil particles were so dispersed and loose that the seeds never got the sufficient soil-to-seed contact needed to maintain the moisture around it to soften and start germinating.

Eventually, as the days pass and your plants are growing, the soil begins to form a hard crust on the surface. Weeds start to rapidly sprout since you brought many dormant weed seeds to the surface with the tiller. Your first inclination is to break out your new tiller and make a pass up and down between the rows to make that soil nice and fluffy again. After all, this is much more fun than the back braking work of hand-pulling weeds or using a hoe. A few weeks later, it's time to do it again and so on, until your soil begins to look permanently different. The weeds have stopped growing-but so have your vegetable plants. Your garden always looks powdery dry so you are watering nearly everyday and your fertilizer is having zero affect on your plants. You finally give up in disgust and break out your tiller again and grind the whole garden up, replant everything and so on it goes. What you have inadvertently done, my friend, is destroyed your soil tilth.

By using a high speed tiller with it's violently rotating tines, you physically changed the structure of your soil. This destroyed in a few weeks what took nature eons to create. Soil can be incredibly complex and fragile. We have different soil types around the country of course, and it is important for us to try to keep what nature provided for us to use, as close to what was there in the first place. There are different ways to accomplish this depending on your soil type. In sandy loam type of

The Gardener's Chairside Reader

soils you can use a hand or wheel hoe to cultivate your soil. The tines of the cultivators will turn over the soil surface rather gently and expose the unwanted weed roots. The weeds will dry up and die, but the basic soil structure remains unchanged. The microscopic clay particles and quartz sand grains are still in a natural ratio to each other. For example, when you grab a handful of soil and squeeze it into a ball it will still stick together. This is important for soil moisture retention. There are many other factors in soil tilth but I chose this example as it's the most common mistake for the average gardener.

Using a tiller is perfectly acceptable, of course, when you have to break up fresh new ground, or incorporate a cover crop into the soil to get ready for a new vegetable crop. But my point is for you to use your tiller judiciously-and only when necessary. A wheel hoe is easy to use even for a chubby old guy like me and honestly, a light till with a hand hoe or garden rake is all you really need for weed control in most gardens. Keep the tiller in the shed and use it only when necessary. Enjoy your soil tilth. You can even brag about your tilth at parties and folks will naturally gravitate around you, buy your drinks and consider you a tilth super hero of sorts. It's a good feeling, too!

Bud Brinkley

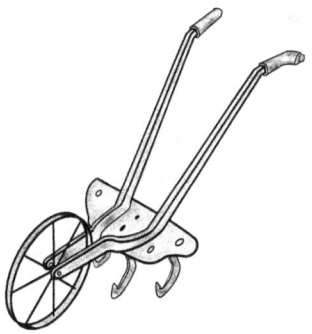

Wheel Hoe Wisdom

The wheel hoe is not a recent innovation. It has been around in many forms for over 100 years and yet, it has stood the test of time. A wheel hoe is a human powered garden implement that consists of a frame with a plow or hoe attached. This frame usually has one or sometimes two wheels attached and a handle bar of some type to push it along through the soil. The real advantage of a wheel hoe is that the weight of the implement is carried by the frame and wheels so it is easy to push while still having precise control.

Early in the century, many variants were designed and sold, but undoubtedly the forward wheel design by the Planet Jr Company proved to be the most popular. Their design moved the wheel well forward of the frame and thus, the forward motion of the operator also kept the plow points buried into the ground with minimal effort. Their unique frame design also made it easy to switch between different implements. Various options from hoes, cultivators, and turning plows were designed for use with this one implement. They even produced a seeder attachment that mounted to the frame to plant seeds evenly spaced at a consistent depth.

Bud Brinkley

When motorized tractors gained popularity around WWI, the wheel hoe faded into relative obscurity until it once again surged in popularity during the 1990's. These days, many savvy gardeners know the value of using a wheel hoe vs. a big and heavy tractor in their small home gardens. A tractor is still a necessity for larger gardens, but a wheel hoe is ideal for a typical home garden. For one thing, the wheel hoe is easy to maneuver, and it does not compact the soil like a heavy and large tractor. Also, a wheel hoe is affordable and easy for a typical gardener to store in their tool shed or garage.

Antique wheel hoes can still be found at barn sales and antique stores and with a little maintenance can still be serviceable for many more years of use. The Planet Jr wheel hoes are also becoming a collector's item, so their prices are usually higher. Several companies are now producing brand new versions of the wheel hoe, and while somewhat expensive, these seem to be well made and designed with versatility in mind.

One drawback of a wheel hoe is that using them in a raised bed type of garden can be cumbersome and problematic. They are designed for smaller garden plots in the ground utilizing rows between the plants. While pushing a wheel hoe is surprisingly easy, it still may be difficult for those who are older or have mobility issues. Lastly, a wheel hoe is only useful up to a certain point. For example, weeding and tilling between individual plant stalks is best accomplished with a hand-held tool such as a hoe or hand rake. Also as the plants mature and fill out in size between the rows the wheel hoe can have difficulty maneuvering around the plants without causing damage to the plants themselves.

Fortunately, antique wheel hoes are still plentiful and are easy for the average gardener to locate, purchase, and use. They make the task of weeding between rows very easy and they do an excellent job of keeping your garden clean and hygienic without destroying your soil tilth. A fifteen minute session with a wheel hoe first thing in the morning a few times a week is all that is needed to keep your garden healthy and looking it's best!

The Gardener's Chairside Reader

Hay Balers and Lands' End

Years ago, I registered for a week-long class at a University nearby to teach me the finer nuances of growing hay and other types of feedstock. When I showed up to the University for the first day of class, the first thing that I noticed was that I was woefully under dressed. I was wearing jeans, a tee shirt, ball cap, and sneakers because well, these are my usual work clothes. Those surrounding me had their brand new straw hats with no signs of sweat stains to be seen anywhere. I also noticed their new fancy leather gloves jauntily sprouting from the back pocket of their new starched khaki cargo pants (you know the style, with the obligatory one hundred accessory pockets sewn everywhere ready to fill with cargo.) All the ladies were displaying the latest new and cute pink rubber gardening boots and one guy was even carrying a very expensive looking leather satchel with the Lands' End logo predominately displayed on the front. This thing looked like something that Indiana Jones would have slung around his shoulder while running out of a cave. The class professor was a suit and tie academic type who had more credentials after his name than the United States Surgeon General. I settled into my chair and he proceeded to start the class with a thirty minute long introduction of himself, I assume to justify why we should believe

Bud Brinkley

everything he was about to teach us.

Admittedly, the class went well this first day as it was all classroom-study and I actually learned quite a bit. On the second day, we were to meet at the livestock barn on the University campus for some field demonstrations. That morning, everyone was still proudly displaying their recently acquired work clothes in a studious manner, and I came dressed a little fancier myself. I decided to wear a freshly washed ball cap and my nasty old rubber boots instead of sneakers.

We were there that morning to watch a demonstration of the newest hay baler machinery in action. We all walked around and marveled at the brand new equipment while the local grad student underlings got everything hooked up and ready to use. Watching all of this fancy equipment in action was fun and interesting. We were walking along beside the baler in the pasture while the graduate underling was driving the tractor. The professor guy was screaming as best he could over the noise of the tractor about the virtues of tight bales when as usually happens, a hay bale suddenly became jammed inside the machine and everything came to a screeching halt. Immediately, the professor barked orders at the grad student on the tractor and the poor kid proceeded to dissect the machine to clear the jammed bale. After a long while of waiting, while standing in the hot sun, the professor decided that we should all walk back to the barn and proceed to the next demonstration. So we left the poor grad student there in the pasture, still digging around inside the stuck baler.

For our next lesson, we gathered inside an open arena near the barn for a demonstration of how to properly take grass samples from a bale of hay to send out for lab analysis. We were fully involved in this delicate operation when the grad student came running and shouting across the pasture from the direction of the tractor and crippled baler. A thick plume of smoke could be seen rising and it was obvious that the baler had caught fire. To make a long story short, several acres of pasture around the University burned up along with a very expensive new baler that day before the local fire department extinguished the

The Gardener's Chairside Reader

flames. It was all very entertaining and exciting. I remember on the drive home that afternoon thinking that maybe I should mount another fire extinguisher on my own baler just in case. So at least I learned that nugget of wisdom.

The remaining courses throughout the week were pretty uneventful and frankly, boring compared to watching the professor run back across the pasture to the towering inferno in his suit with his tie flapping about his neck. By the end of the week, about half of the cargo pants crowd had dropped out and all that remained was me, the Lands' End dude, and a few others. I sure wanted to get a peek inside that satchel he was still carrying everywhere. I may have to get one of those for myself someday…

Bud Brinkley

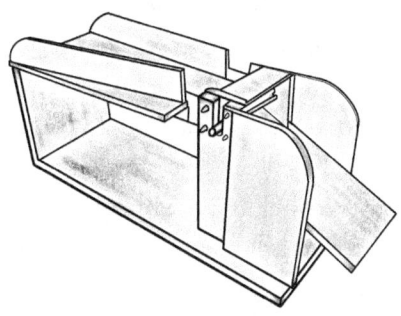

The Brinkley Shell-O-Matic

Unlike most country folk, I hate shelling peas. There, I said it. Sitting around shelling peas all day is my idea of inhumane punishment. Most likely this came from being forced to sit and shell buckets of peas when I was a squirmy and hyper-active 8 year old child. If you do it long enough your fingertips get raw. Eventually just seeing all those peas ready to pick on the plants makes you avoid picking them with just the slightest excuse. That's my standard go-to method of avoiding the work of pea-shelling.

My habit of watching YouTube videos showed how easily the big commercial electric pea sheller machines could do the work for you. A guy dumps in a bushel of peas into the top-hopper, and in just a few minutes beautiful clean little peas emerge in a tub ready to eat. What a dream! That's what I wanted, but being the tightwad that I am, there is no way I was going to purchase one of those. In desperation, I did buy a small hand sheller that you clamp to a table top and turn a crank while feeding it one pea pod at a time. An improvement for sure, but this thing was built too light for even medium duty usage. My engineering mind set about to improve it and as a result, I built my own electric pea sheller.

Bud Brinkley

I am an accomplished woodworker and I have a nice workshop equipped to build everything from wooden boats to furniture, so it was a natural progression for me to build a wooden pea sheller also. I carefully examined the hand crank sheller that I had purchased and set about studying how it worked and how I could improve it. This sheller design uses two small plastic rollers that pull the pea pod through the rollers and splitting the pea shell open while doing it. The empty pea shell passes between the rollers while the peas fall into a container below. The whole process is very much like the wooden ringers used on the old washing machines. It all looked simple enough, so I tore the flimsy hand crank sheller completely apart and started to work, building a new and improved pea sheller.

After a long day of sawing, gluing, and sanding I had in front of me a fancy motorized version of a pea sheller. The new design included sturdy wooden frame and used a cordless electric drill for power. I devised a little guide chute to drop the newly liberated peas into a plastic bowl while the empty pods slid down another chute into a cardboard box. It all looked spiffy and seemed to be exactly what I needed to get several bushels of peas shelled in a short time with minimal effort. The best part would be no raw fingertips! Finally, I was ready for a test run.

I picked a few pea pods from my garden and set up my new invention on the workbench, adjusting the rollers. I squeezed the trigger on the electric drill motor and the whole thing started whirring and shaking a little, but it seemed ready for action. I inserted a single pea pod and the machine started crunching on it while pulling it through the rollers. The flattened pea pod slid down the guide chute into the cardboard box just as I had intended and I looked into the plastic bowl below and I saw 10 beautiful little peas! Eureka! I quickly ran a few more pods through the little machine with the same good results. Feeling confident after a bragging show-and-tell session to my wife demonstrating my newest invention, I ran back to the garden to grab up an entire bushel of purple hull peas.

The Gardener's Chairside Reader

I let the peas hulls dry a bit overnight before trying to shell them as it seemed to work better that way. The next morning, I could hardly finish drinking my coffee in eager anticipation of what would be my first work day with my new sheller. I started shelling the peas slowly at first. I eventually found a rhythm and I was shelling peas at a good clip. That morning I had a gallon of shelled peas ready in about 5 minutes. I think I ended up shelling around 4 bushels of peas with great success. A relative who happened to be visiting us at that time also marveled at how well the little machine worked. I think he called me Thomas Edison or something, and it made me real proud.

I have been happily using my pea sheller for several weeks now and at the end of each day I carefully wipe it clean and polish and oil the rollers. In my mind it's a precision piece of equipment that deserves the lavish care that I shower upon it. I named it the Brinkley Shell-o-Matic' and it has earned it's rightful place of honor upon my shelf in the storage closet, ready for action at a moment's notice. I have now graduated one step above all those poor and lowly mortals who have to sit in their rocking chairs with their raw fingers and their tiny little bowl of peas in their lap. No longer do I have to chain myself kicking and screaming to the back porch with a bushel of peas waiting at my feet. All I have to do now is squeeze the trigger on the drill and watch those beautiful little black eyed peas fall into their bowl and laugh like a mad scientist in those old horror movies.

Bud Brinkley

The Gardener's Chairside Reader

Lunar Post Holes

I'm not much on believing various superstitions when it comes to gardening. Things like planting by the moon or dancing over your squash plants just don't sound like something I would try. I am a scientific type of guy most of the time and I try to find some reasoning for these things instead of blindly accepting old myths and legends. Still, it's fun to talk with folks and listen to their tales. I find it interesting too sometimes.

One morning, I got talking with some utility guys that were digging a ditch up by the road. They told me that they always tried to schedule their digging work on or close to a full moon. I asked why that would be and they said so they could be sure to have enough dirt to completely fill their trench when they are finished working. They told me that if you dig a hole or a ditch on a full moon you will always have a too much dirt left over when you refill the hole. If you do the opposite and dig during the dark moon you'll always be a little short. I laughed as that's the first time that I heard that tale. They all seriously swore that it was true. After all, these guys dug ditches for a living so they must know what they were talking about. I was intrigued.

Bud Brinkley

Since it was a full moon, I decided to experiment on my own. I took my post hole digger and dug a 2 foot deep hole in the ground and I also dug a short section of ditch beside it. The next day I returned to cover it up and by golly sure enough they were right! I had excess dirt from both holes. I then waited a few weeks till the moon was as it's darkest cycle and tried again. Yep, the next day I did not have enough dirt to completely fill the holes.

I researched online a bit and the theory is that the lunar gravitational pull is stronger during a full moon, so this must have some effect on the soil as you dig it up out of the ground and perhaps loosens it a bit and expands it. When you try to replace the dirt there is now a little too much soil left over to fit in the hole. Even though I can see the scientific reasoning and possible explanation for this phenomenon, I still stubbornly reject most other old tales. However, I am tempted to try dancing over my poor squash crop this year to see if that does anything.

The Gardener's Chairside Reader

Christmas Pork Chop

One year around Christmas time, I was tasked by my wife with decorating our ranch entrance gate with some Christmas decorations. This is commonly done in the country and it's always fun to see the creative ways that rural folks decorate their gates. My wife was busy working long hours at her job so I volunteered to drive to town to purchase some new decorations as our old ones that we had been using in the past were all just about worn out. I went to a large town nearby to the local crafts store to do some shopping that morning and unwittingly set forth a chain of events that is forever burned into my mind.

Shopping is not one of my favorite activities and I feel uneasy sometimes walking into the feminine domain of something like the craft store Christmas decoration department. I was wandering around the isles by myself that morning aimlessly trying to decide what to buy, when a lovely young lady walked up pushing her basket and said, "You look very lost. Can I help you with anything?" Now one of my more annoying traits is that I tend to talk with a very loud voice and my wife is always telling me to talk quieter in public. I must have blurted out something like, "I'm trying to find some Christmas decorations to decorate my ranch gate."

Immediately, every woman in that store heard me and their radar switched on. All of these lovely suburban women immediately envisioned I was a rich, single cowboy who owned a big cattle ranch and I desperately needed their help, of course! Women came running from every aisle and shopping baskets were beginning to collide as they pushed and shoved their way over to me. Decorations were beginning to be thrown into my shopping basket and garland was tossed into the air. I had about 10 women all talking to me at once asking me questions that I shall not repeat. The overwhelming scent of 10 different types of perfumes was more than I could handle. I now know exactly how a pork chop feels when it's thrown into a pen of hungry wolves. They thought they all had found the jackpot that day when in reality, I was just a dumb old married husband simply trying to help out his wife with some chores. I eventually made my way out of the store and back home. I'm not sure what all I had ended up buying that day, as some of those decorations are definitely not something I would have picked out. I told my wife that evening what had happened and she just shook her head and rolled her eyes at me.

We laugh at that story each Christmas. You can bet that I stay out of the craft store Christmas decoration department these days and my wife now keeps my loud mouth taped shut too.

Farm Gate Envy

I have always had a somewhat strange obsession with farm and ranch gates. It probably started at an early age playing on our old wooden gate that my dad had built on our farm. It was huge and heavy and painted bright orange. Later in my life while working in the big ranch country in South Texas, I observed and studied many ranch gates. These gates guarded the entrances to some very large ranches in the area. To give you an idea of the size and wealth of these ranches, you drive through the entrance gate by the road, and then you may drive another 50 miles *inside* the ranch before you came to the actual residence! Some ranches had their entrance gates adorned with huge, and beautiful rock columns, elaborate water fountains, and fancy custom iron work. Others were just plain and simple everyday gates even though the ranch probably had a landing strip and a private jet in the hangar next to the house! I suspect the owner's ego had much to do with the design. Studying the varied gates through the years inspired me to decide between form and function when it came time to build my own gate.

At our ranch entrance, I opted for a simple gate using wooden fence boards around the entrance and a 16ft. metal tube gate that I painted

Bud Brinkley

green. Through years of weathering and rusting, the bright green has now turned a grayish mossy green but I still rather like the color anyway. When I built our gate I knew we would be coming and going through it in all kinds of weather so I installed a fancy new electric gate opener. This opener was a pricey extravagance for us, but nothing is too good for my sweet wife, so it was a convenient necessity.

For privacy, we built our home on the backside of the ranch, so it is exactly 1/2 mile from the main gate to our house. We also have a smaller, secondary gate near the house to keep the cows out of our yard. The end result is that on a good day it takes about 10 minutes to drive from our house to the main road after passing through both gates, dodging Olympic pool sized pot holes, and driving around the stubborn cows and their poop in the pasture, etc.

This arrangement worked nicely for several years until my Father-in-Law used my gate one day to bring in some heavy equipment and his trailer hit our electric gate opener smashing it. (Back story: I have always said that man could tear up a metal anvil using just a toothpick.) After he "fixed" the gate for me, things just never worked the same. One rainy morning, my wife was leaving for work dressed in her fancy clothes and nice sandals. The first electric gate opener had a dead battery so she had to get out of the car in the rain and manually open the gate, get back into the car and drive through, get back out of the car into the rain again and shut the gate, get back into the car, and proceed to drive across the soggy pasture to the front gate only to realize the front gate wasn't working also! As a result, her clothes were now soaking wet and she had sticky mud and probably cow manure oozing up between her perfectly manicured toes. A few minutes later I received a text from her on my phone that I will only classify as - charming and descriptive.

Twenty years and five sets of replacement gate openers later, I still have to get out and manually open the dang gates. I have learned a few tips along the way. Tip #1 - on rainy days I try to ride ahead of my wife on the 4 wheeler and open the gates for her, or else I will face the full blunt force of her disgust with oozing toes. Tip #2 - if you need to

The Gardener's Chairside Reader

invite guests or service people to your home, meet them at the front gate to let them in as they will never get the thing to open properly, or they will end up getting lost on your ranch and probably get their shiny new car stuck in a hog wallow someplace. Tip # 3 - always keep your gates closed to keep out unwanted relatives, salesmen, and nosy tax appraisers. It has often been said that good fences make good neighbors. I think fancy electric gates keep Mama happy, and that keeps everyone happy.

Bud Brinkley

The Gardener's Chairside Reader

Salt Of the Earth

People who grew up in America during the Great Depression of the 1920's were truly the "Greatest Generation". They really knew what it meant to be poor, hungry, and they knew how to find hope during a hopeless time. History has shown that this generation all worked together collectively to pull this great country back together. I was extremely fortunate as a young kid to grow up knowing two such wonderfully resourceful people, Mr. And Mrs. Murphy. The Murphy's were a part of my life from my early childhood till I was a grown man when they passed. What they taught me by example planted a seed of compassion that I try to have for those around me.

My Dad first encountered the Murphy's when he was a young man himself in his 30's. He was hunting early one morning and not having much luck. He stumbled upon an ancient old school bus parked on a derelict homestead. He said it looked as though it had been abandoned for many years, and so with my Dad being somewhat mischievous, he decided to have some fun and try a little target practice on the old bus. It was a different era back then and it was somewhat commonplace to pull such a stunt. He squatted down in the fog and took aim at one of

Bud Brinkley

the windows on the bus and fired a shot right through the glass. He heard a commotion and suddenly a skinny fella burst out of the bus still in his underwear yelling, "Don't shoot!" My Dad jumped up out of the bushes and apologized profusely telling him he didn't realize anyone was inside. It turned out that Mr. Murphy and his wife were living in that bus! They were inside and asleep that morning when Pops fired the fateful shot. Thankfully no one was hurt and my Dad promptly returned home, gathered up some tools and supplies and returned to repair the damage he caused to their bus home. That day was the beginning of a long and wonderful relationship my family had with the Murphy's.

Mr. Murphy was a farmer and a logger who cut trees for a living. He scratched out a living during those lean times doing what ever he could find to feed his family. After their children grew up and left home they sunk back into poverty and ended up somehow living in that old bus. With the help of my Dad they eventually built a ramshackle farmhouse and moved out of the bus a few years later. They moved around to various places several times over the years and each time they moved, they would pack up their junk and move that old farmhouse with them to their new property using his log truck. I last visited them one day about 25 years ago and they were still sitting on the rotten porch of that little house. Mr. Murphy was always a treasure trove of farming advice and his sweet old wife was the type who could butcher a hog at daylight and have it ready for supper that evening and she did it time and time again.

Oddly, another hunting story comes to mind that connects our families together. One evening Dad and I went squirrel hunting on a cold winter day in the national forest nearby and somehow we became separated from each other and lost at dark. I eventually found my way amongst the black woods back to where we had parked our truck and crawled into the cab and waited for Dad to return. I nearly froze to death in that pickup waiting all night as Dad never showed up. The next morning at daylight, I knew Dad was in real trouble so I abandoned the safety of the truck and started walking out of the forest for help as I was too young to know how to drive yet. I had walked about 5 miles or so through the deep woods when suddenly I saw a rickety old log truck rumbling up a makeshift logging road. Sure enough, it was Mr. Murphy about to start his day's work cutting trees. I ran up to his truck and he was as surprised to see me alone in the woods as I was him. I explained what had happened to me and Dad, and he immediately took me back home to their little cabin. Mrs. Murphy hurriedly fed me some lard biscuits while Mr. Murphy grabbed his gun and hound dogs and we headed back to Pop's truck that I had previously abandoned to search for Dad. Fortunately, all ended well. Dad had eventually found his way through the forest early that morning and he was sitting in the pickup, looking very worried and somewhat disoriented when we returned to find him. He was scratched up, cold, and hungry but I suspect what he feared most of all was my Mother's wrath when we would finally arrive back home. That wasn't to be the last time Mr. Murphy saved me.

Years later, my father died when I was in my early 20's. After that, I somehow just lost my way in life as a young man. By the time I was in my 30's I had encountered an unfortunate series of events that ended in a divorce and subsequent loss of my business and home. I had hit absolute rock bottom, sunk into a terrible depression, and literally the only thing I owned were the clothes that I had on. I still had a lot of debt and bills to pay and my bank was hounding me to come up with something, anything to pay off my business loans that had gone into default. I was living in a small town with a small bank, and gossip quickly gets around about stuff like that. I had borrowed my Mother's old car and planned to drive home as I had no remaining friends around there, and I had nowhere to stay either.

As I was driving out of that wretched little town for the last time with my tail between my legs, I noticed an old rickety log truck rumbling up the road passing me by in the opposite direction as it headed back towards town. It was Mr. Murphy and his wife, who was bouncing along in the seat beside him. I was relieved that they did not recognize me in my fog of shame as we passed each other and I continued to drive home to Mom's house feeling defeated. I found out a few days later that they had heard the gossip about what I was going through. When we

Bud Brinkley

passed each other on the road, they were driving to town that morning to go to my bank and they quietly made a deposit into my account. They had scratched together what money they had saved, and they chose to use it to help me get back on my feet again.

That was many years ago, and since that time the Lord has blessed me beyond measure. I now have a wonderful life, prosperity, and a truly loving wife. Sometimes I sit quietly and think about the Murphy's and all that they represented as good, solid country people. I get teary eyed too, and will be eternally grateful for what they did for me that day and for being close to my family through the years. May God give rest to their gentle souls.

The Gardener's Chairside Reader

A Quiet Morning With Big Bertha

Mowing grass is never-ending on a ranch. One day I needed to get my big tractor out to mow some clearings around the property. This tractor is very large, somewhat old, and has an enclosed cab in which to sit and drive the beast. It does not get used much. I don't use this tractor very often because by the time that I repair all of the the hydraulic leaks, go to town to buy a new battery, pump 25 gallons of diesel in the fuel tank and all the rest of the things necessary just to get it cranked up, I'm usually dead tired before even beginning to do any work with it. My old tractor sure had a surprise waiting for me that day.

"Big Bertha," as I call her had been sitting around all winter in a remote part of my property. When I used it last season, I had run out of fuel which is a troublesome deal with diesel engines. I decided to just let her sit out the winter right there in that spot and figured I would get her running again next spring. Well spring was here, and bushes and weeds were sprouting everywhere, so it was time to wake up' ole Bertha from her slumber. The ole gal fired right up after a bit of coaxing and praying and I crawled up into the cab to settle in for a day's work.

Bud Brinkley

It was loud as heck in the cab, so I had a habit of wearing a pair of thick noise-canceling headphones to protect my hearing. When wearing these headphones you become somewhat mentally detached from the machinery that you are operating. This can be good or bad depending if the tractor's engine is making strange new noises, but I'll leave that tale for another story. Anyway, I had been in the noisy cab running the tractor all morning for about three hours. I was getting thirsty, so I was shuffling and twisting around in the seat while I was still driving along trying to get my Thermos open for a drink. Something suddenly moved under my seat and caught my eye as I was pouring some coffee. I looked down between my legs and there was a huge black snake coiled up in the cab with me next to the toolbox and my foot!

Now this was no ordinary small grass snake. In the one millionth of a second that it took for my brain to process all that was happening here, I imagined that it was about 6 feet long and about as thick as my arm. It had devilish eyes that pierced through it's prey and paralyzed them in fear. In the very next one millionth of a second I reached over to open the door of the cab and I bailed out. Yep, I jumped out of a running tractor as there was no way that I was going to share that tiny space with a viper who could eat a small deer for lunch. Now jumping out of a moving tractor is inherently dangerous, but that did not stop me. I was getting out of there and I would just have to take my chances.

I hit the ground with a painful thud and sprung to my feet. I spit the dirt out of my mouth and started chasing the now driver-less Big Bertha as she was still lumbering along dutifully mowing a neat swath behind her with the big brushog mower. A large pine tree was in her path and the two met with a thundering crash. Ole Bertha was not about to give up and the big old pine tree had no plans for budging either. The tractor just sat there growling mightily and spinning her giant wheels in the dirt while pounding her nose against the tree trunk making all kinds of racket. Meanwhile the snake decided he had enough of the jostling around and casually slithered out the open door of the cab and down the steps of the tractor and finally made his way to safety into the woods. It was a huge relief to me to see him crawl away and I managed

The Gardener's Chairside Reader

to get back into the cab without getting run over and got Bertha shut down and all was quiet for a few moments while I stopped shaking and gathered my senses.

I subsequently performed a detailed microscopic inspection of every nook and cranny of the tractor cab for anymore snakes and then I finished up mowing for the rest of the morning. Today, once again Bertha is silent and sitting by herself peacefully till another season. Only this time she has her door and every single little hole and all of the various openings in the cab tightly taped shut with duct tape.

Bud Brinkley

The Florida Weave

Many vegetables that we grow in our gardens need a trellis. This is usually some sort of supporting structure that either keeps a heavy plant from collapsing to the ground or it gives a naturally climbing plant a surface to cling to and try to climb and spread out. Trellises can be made from just about anything and have probably all been documented on YouTube at one point or another. I have watched people make a trellis with everything from bamboo to heavy scrap iron beams. One time I thoroughly enjoyed watching some woman make a huge trellis from just stalks of dried grass stems. She was a self professed expert and she worked for probably a full day on this one trellis. It was hilarious because the wind kept blowing her towering trellis apart before she could construct it. Finally in frustration she mentioned that it would be best to construct it on a calm day. My first thought was what about those windy days after you have built this monstrosity? Tomato trellises are the thing most sought after by the average backyard gardener and so I want to pass along to you a handy design that I learned how to construct from you guessed it - YouTube.

Somewhere in time, a very smart person designed a trellis system using metal fence posts and twine. I'm left to assume this person was

Bud Brinkley

from Florida because they proudly named it the "Florida Weave Method." This method of trellising tomatoes is used by the large commercial tomato farms around there as it really works best for plants that are planted in rows. Short rows will work with this method, but your plants must be planted in rows. The basic design is to place metal fence posts about 10 feet apart along the centerline of the row, between the plants. Next, take some twine and tie it securely to the first post. Holding the ball of twine in your hands, proceed down the row weaving in and out of the plant stalks till you reach another post. You tie the twine to the next post and then continue the pattern until you reach the end of the row. At that point, you work your way back up to the beginning of the row weaving and tying in the opposite direction. When you are finished you have a long run of twine that is woven into the plants on both sides of the plant stalk, thus securely supporting it. The whole process is actually very easy and it supports the plants very well in all kinds of wind and weather. As the plants continue to grow taller you just repeat the process periodically by tying a new row of the twine about a foot or so above the previous run.

I tried this method of trellising on my tomatoes this year for the first time. It worked very nicely with a few minor adjustments. My tomato plants were very heavy because I had a great crop this year. The weight of the giant green tomatoes was too much for the lightweight bio-degradable twine that I initially used. When the twine got wet in the rain or a heavy morning dew, it would stretch out and sag, so I ended up using some hay baler twine like that used to tie the big round hay bales. This worked very well for me. Also, due to the large size of my plants this year, I think it would have been much sturdier if I had spaced my metal fence posts at 8ft. instead of 10 ft. apart. Overall, this method works well for any tall and leggy type of plant and I have used it on my pepper plants with success also.

Try the "Florida Weave" method for yourself sometime and I'm hopeful that you will have the same success.

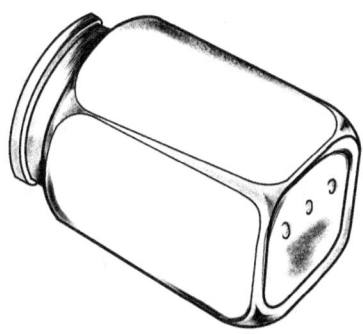

Three Dot Snuff, Cigars, and Old Dogs

One day my Dad and I were headed out fishing and, as per our normal routine, we first made a stop at an old country general store to purchase some worms for our fish bait. Well, that's what I was told each time, but I concluded as I got older that Dad really wanted to stop and buy a Seven Up soft drink to mix up a little toddy with his whiskey. Dad loved Old Crow whiskey and since Mom usually wasn't with us on these fishing trips it was the perfect opportunity for him to mix up his favorite drink of choice. We pulled up to the store in our truck and walked inside.

These stores always seemed to look the same no matter where we stopped. We walked across the ancient wood floor that had long settled into the ground and it was a bit like trying to walk on water. We hung around the place while Dad talked with the store proprietor-mainly because back then the soft drink drink bottles had a stiff deposit on them so we had to finish the soda and then return it to the store before leaving to avoid paying that deposit.

On this particular day in the store, I was chugging away at my half of the Seven-Up when I noticed a giant old dog laying smack in the

Bud Brinkley

middle of the store on the floor gasping for each breath in the sultry heat while he slept soundly. Suddenly, he awoke from his slumber, painfully got up, scratched his giant old ears, and lumbered over to the dusty old glass display cabinet next to the cash register and started whining and barking his raspy old-dog voice.

These type of country dogs are always similar in that they were very old and fat and had droopy eyes. Their ears were floppy and scarred from the years of scratching them because of the fleas. You could smell them from about 10 feet away too. The store owner who was still talking with my Dad, casually reached down and opened the display case and reached into a box of King Edward cigars and gave one to the old dog! The dog happily chomped it in his mouth and walked back to his sleeping spot on the floor and ate the whole cigar before laying back down to resume his nap. My Dad chuckled and ask the store owner about that and this is the story that he told us...

In his younger days, the dog was used as a guard dog inside the store. They would lock him inside the store at night to deter any would-be burglars and such. One night, the mischievous dog broke into the display case and ate an entire box of King Edward cigars and in the process, he got addicted to those cigars. The store owner said that the old dog was particularly keen for King Edward's and he would not touch any other brand of cigars. Ever since and for many years afterward the old dog would hobble over to the glass case every few hours and cry for a cigar to chew on.

On another fishing trip, in another country store, Dad and I were doing the exact same Seven-Up ritual again. I was choking down a half bottle of Seven-Up as fast as I could so Dad could get back to his whiskey concoction waiting outside in the truck. Three old ladies came into the store. All these ladies were dressed in their usual daily uniforms consisting of a long skirt, an apron, and and a sun bonnet or some old rag wrapped around their head. They too were going fishing for the day as they all sported homemade cane poles. They all laughed loudly and were very jovial and kind to me as they walked past. Apparently they

The Gardener's Chairside Reader

wanted to buy some snuff before heading down to the creek bank under the road bridge nearby. They each asked for a bottle of "three-dot-snuff." The store owner turned around to the shelf behind the counter and looked amongst the rows of snuff jars neatly displayed along the shelves. These snuff jars were brown in color and square shaped. They had an ancient, zinc screw-top lid and inside was the powdery snuff. As the store owner turned each jar upside down and looked underneath, he would look for the jars that had three dots on the bottom. Some jars had two dots, some had three dots, and some had more. Once satisfied that each lady had the right dots on their jars he tallied up their charge accounts and they left with their snuff for the day. I asked my Dad about this and he explained what I just witnessed.

On the bottom of the snuff jar the manufacturer usually had the name of their company molded into the jar along with the dots or bumps just under the logo. Countless old-timers came to believe that the jars with the three dots contained stronger snuff than the ones with two dots. Four dots and above was really strong snuff and probably reserved for the old men and grizzly farmers who could handle it. In reality, the dots were just a manufacturing code that the jar manufacturer used to identify which factory molded the jars much the same as modern barcodes are used today. The snuff inside all the jars was exactly the same strength. But these old ladies adamantly refused to accept that explanation and they all swore that the dots indicated the strength of the snuff. Nothing was going to ever change their minds ether.

I'm sure those three old women sitting on their buckets next to the creek bank with a cane pole in their hands and their three dot snuff in their laps probably out-fished me and Dad that day. I'm also sure that old smelly dog probably made sure that his owner kept his personal supply of King Edwards in stock at all times too.

Bud Brinkley

The Gardener's Chairside Reader

Suspended Animation

For most of my life, I have been a jeans and t-shirt type of guy. These have been my standard attire during the mild winters we have around here. I live in shorts for the summer, but that's getting more embarrassing each year as I think old guys look funny in shorts and I may have to modestly curtail that practice in a few more years. For example, there is an older gentleman who lives nearby and he always takes several daily trips walking up and down the road. Unfortunately for everyone who sees him, he always wears those tight gym shorts that were popular back in the 1980's. You remember those. They were constructed of a kinda shiny type of fabric and were very short. They had the scalloped leg cutouts with the racing stripes sewn into each side. I suppose in his mind those kind of shorts are still the current fashion and well-made, because he has been wearing the same pair of gym shorts for probably the last 40 years. One day several years ago while driving my elderly mother to town, we passed this man on his daily walk. When my mother saw him dressed like that, in his short disco shorts, she shook her head in disgust and said a few caustic remarks about his physique. I looked at her in horror and laughed, but I took a mental note that maybe I should be a little more careful about what I wear also. Incidentally, he is still walking up and down the road,

Bud Brinkley

but only now he is wearing sweatpants. I suspect somewhere along the way my mother may have got a hold of him and told him how the cow ate the cabbage about his fashion faux pas.

I am starting to sympathize with the old man in the shorts as I get older. Wearing a pair of jeans is getting almost impossible for me lately. I hate shopping for clothes with a passion, mainly because all of the store racks these days are full of clothes for young 20-somethings with a 28" waistline. Old fat dudes like me have to sheepishly dig around in the "big and tall" section that is usually well-hidden near the clearance rack. I wished they had a "short legs and fat belly" department as that would make things much easier for me, for sure.

I noticed on TV one day while watching an old movie that this fella had a manly physique much like me and his pants also were taking on that all too familiar "belly dive" around his waste. To compensate and to keep his pants up, he was wearing suspenders. I liked that simple solution so I assigned my sweet wife the task of buying a pair of suspenders for my work jeans. A few days and $300 later (she said she needed a few things herself) she came home with a nice pair of red suspenders. I looked suspiciously at their heavy duty construction and the large clips to attach them to my pants. I knew at that point a major life milestone was about to take place.

At first, I could not get my new suspenders to attach properly to my pants. It turns out that I am too chubby to reach behind me to clip them on. I devised a clever method of clipping them to the back of my trousers before I put them on. After that fiasco was settled, I had to adjust them a bit. The first time I tried them on they were way too short and they pulled my jeans hereto into unexplored territory, shall we say. After my voice returned to it's natural pitch and tone I realized the suspenders were actually comfortable!

These days I will not wear a pair of jeans without my nifty suspenders and I own several pairs in different colors to suit my mood. Whenever I do manual work outside in the garden it's a great feeling to not worry about my belly pushing my pants down to my knees every few minutes. I suppose that someday when I really am getting old that I will have to abandon the jeans/suspenders deal and go into full stage-3 mode with wearing overalls and a denim shirt. But for now I will just keep snapping and buckling each morning. I still have those old knee high white tube socks with the three brightly colored stripes at the top though. I ain't giving those babies up yet.

Bud Brinkley

The Gardener's Chairside Reader

A Honeybee's Quest

The honeybee is, without a doubt, the hardest worker in the garden. You and I cannot possibly compete our daily labors to enormous tasks that the bee accomplishes with clockwork precision. Honeybees are quite simply, fascinating to study. Most bees live together in colonies that have a hierarchal social order and each colony consists of several different classifications of workers. All of these colonies usually contain thousands of bees, the majority of which are worker bees. These important members of the colony are the ones who travel away from the colony far afield in search of nectar and pollen. Once they have filled their honey sacs with a load of pollen, they then proceed back to the colony to distribute it inside the hive. Each worker has a specific task and interestingly, these tasks may even change as the bee gets older and matures.

Honeybees can and will travel great distances to look for the choicest flowers. Many times a bee may travel several miles if the blooms are scarce or if other man made conditions interfere with their foraging. Most bees live in wild colonies that have been established in hollow tree trunks, old structures, or even in a hole in the ground around an old tree stump. Most folks recognize the ubiquitous bee hives stacked in a field or clearing. These hives contain domesticated bees. Farmers and professional bee-keepers move these hives throughout the flowering season so that they will pollinate different flowers. These different flowers will cause slightly different flavors in the honey.

As far as gardening is concerned, the worker bee that is also known as a forager is of utmost importance. Foragers are almost always female and they are the honeybees that we typically see in a garden flying from bloom to bloom gathering nectar and pollen. While working at this task, the bee also manages to cross pollinate flowers, a vital part of vegetable production. By spreading pollen from a male flower to a female flower, the plant is then able to produce fruit. Other insects can also cross pollinate flowers in the garden, but their efforts pale in comparison to the bee.

Since bees are important for some vegetable crops, it is advantageous to place a man-made hive nearby to entice the worker bees to visit the garden area. Usually a hive is placed about 100 feet or so from the main garden and this is close enough to attract the bees without becoming a safety hazard. Sometimes a gardener is fortunate enough to have a wild colony of bees nearby and so a hive box is unnecessary. As soon as the vegetable plants start to bloom the gardener should take notice and check for bees. If many bees are prevalent, there is probably a natural colony within easy distance to the garden and so no hive boxes are necessary. If the bees are absent or noticeably low numbers are present, the gardener will want to bring in a beehive and set it up nearby if possible.

The unrestricted use of pesticides in the garden can have a devastating effect on bee populations. Using certain pesticides can quickly kill an entire colony. Bees are usually the most active in the early mornings as this is when the vegetable flowers first open their blooms. Many flowers close or even drop off from the plant after a few hours, so the bees are naturally more active in the morning and thus human activity around the garden needs to be curtailed at this time. Many folks

The Gardener's Chairside Reader

water their plants in the morning, but this can make it difficult for the bees to do their work. If a pesticide is needed on the plants, it is always a good practice to apply it late in the afternoon or evening so that the bees are less directly affected.

Bee colonies are somewhat transient in nature, so a colony may swarm and move occasionally. The gardener should always take note of their bee population and be prepared to add beehives if necessary. Stepping into the garden in the early morning to closely examine a bee collecting pollen can be fascinating and fun. You can see their little sacs on their legs full of pollen and you can't help but wonder how far she has to fly back to her hive. Most bees are cautious and gentle in their nature so it's fun and easy to get a close look at them performing a miracle right before your eyes!

Bud Brinkley

The Gardener's Chairside Reader

Grace

So many folks take so much for granted these days, that they lose sight of the blessings of God's earth beneath their feet.

Many years ago, I worked in the oilfield industry. At one point, I found myself sleeping on a picnic table outside a locked-up heliport in Louisiana. I was waiting for the 4:00AM helicopter to take me offshore to a drilling rig nearly 100 miles out in the Gulf of Mexico. After landing on the offshore platform, I slowly made my way downstairs to the crew quarters on the drilling rig, and settled in for yet another long two-week "hitch." After two weeks at sea you become accustomed to the noisy sights and sounds of a huge drilling rig. The howl of the giant diesel engines roaring incessantly night and day is unrelenting and your brain eventually just turns it off. The rigs are almost always painted gray. There is no color on a drilling rig to speak of, just gray everywhere. There is nothing attractive or fun or enjoyable on a rig. It's built for hard labor and so it is a strictly utilitarian environment. You can be in very close quarters with 100 other men and women 24 hours a day, yet you are the loneliest that you'll ever be in your life. I did enjoy the work

Bud Brinkley

aspect of my time offshore, but the total disconnect from the rest of my world back on land was something I'll always remember dreading while leaving home for my work.

I was always a bit startled each time I flew back to land at the end of my two week hitch. I could still hear my ears ringing as the helicopter shut down and the rotor blades slowly come to a halt. I sat inside my car in the parking lot of the heliport and immediately I was consumed by the silence surrounding me. Beautiful, quiet, silence. While driving home to Texas, I was always enjoyed the scenery along the way. Trees! Real green leaves and grass! The smell of the air on land is totally different than at sea. I could go on and on…

That was a long time ago for me, but it forever changed how I see my surroundings. Now each day I walk outside and smell the ozone in the fresh morning air. I watch the birds singing and take a few moments to play with my cats. As I make my way to my garden and cabin to have a cup of coffee and plan my day's chores, I notice the weather, and my surroundings down to the last detail. You see, these are all God's blessings and I do not want to overlook a single one. Life can be wonderful for anyone who will take the time to notice the little things surrounding them.

No matter where you find yourself today, you can always rely on God's grace to carry you through the tough times. We've all had them, and with His guidance, we will all overcome the challenges of life that sometimes step in front of us. Humble prayer will do that for all of us.

Pray my friend. Pray each and every day. God Bless.

The Gardener's Chairside Reader

THE END

About the Author

Bud Brinkley has been gardening for over 50 years in his home state of Texas. In that time he has perfected his techniques and shares his knowledge.

Growing up in a rural setting has allowed him to soak in the rich cultural environment and the sometimes humorous anecdotes that living in the country provides. He has a wit and wisdom of all things gardening and he loves to tell a story to his readers.

Since his retirement in 2017, Bud has embarked upon his lifelong dream to live simply, write often, and enjoy life.

Connect with me online:

bud@budbrinkley.com
www.budbrinkley.com

www.ingramcontent.com/pod-product-compliance
Lightning Source LLC
Chambersburg PA
CBHW071403290426
44108CB00014B/1670